Technological collaboration in industry

Technological collaboration in industry

Strategy, policy and internationalization in innovation

Mark Dodgson

London and New York

For Jack

First published 1993
by Routledge
11 New Fetter Lane, London EC4P 4EE

Simultaneously published in the USA and Canada
by Routledge
29 West 35th Street, New York, NY 10001

© 1993 Mark Dodgson

Typeset in Times by Michael Mepham, Frome, Somerset
Printed and bound in Great Britain by
Mackays of Chatham PLC, Chatham, Kent

British Library Cataloguing in Publication Data
A catalogue record for this book is available from the British Library.
  ISBN 0–415–08230–7

Library of Congress Cataloging-in-Publication Data has been applied for.
  ISBN 0–415–08230–7

Contents

Figures and tables

Preface

Technological collaboration in industry occurs for a wide variety of reasons and it manifests itself in a number of forms. It has stimulated an enormous range of interest from academics and industrial analysts. A great deal of this interest has focused on specific forms of collaboration, be it joint ventures, or research consortia or user/supplier links. It uses specific tools and language of analysis, be it from economics or management and innovation studies or organizational behaviour. This book, although not going into the depth of some of these individual analyses, will try and present a broad picture of technological collaboration and, using some of the wide range of tools and analyses, provide a more complete view of the 'collaboration phenomenon'. It is necessary to do this as collaboration is such a multi-faceted activity, and explanations for its multiplicity of motives, processes and outcomes cannot be anything but multi-disciplinary.

Technological collaboration occurs essentially because of the complexity and uncertainty of technological innovation. It is concerned with capability building and learning. Technology is rarely created and marketed entirely by means of the actions of brilliant individual scientists, engineers or entrepreneurs, or through affluent and well-organized research groups in individual firms. Instead, its initiation, formulation and diffusion depends on complex interactions between individuals and groups of people in the science-base and research organizations, firms acting as vendors, customers, partners and competitors, and the changing demands of governments and individuals as customers and regulators. The necessity of linkages between these actors in order to develop new products and technologies and access new capabilities, and the maze of forms they take, inevitably leads to some elements of formalization as firms try to control the complication of it all. These formalized links are here called collaboration, and reflecting the whole complexity of the relationships, the differing motives for them, and wide-ranging forms and outcomes, a broad range of collaborations are considered.

This book provides a review of some of the major issues of technological collaboration: its extent and form; explanations both theoretical and practical for its existence; and its significance for firms, governments and the development and diffusion of technology. The issues are not clear-cut. Evidence on, and interpretation of, the collaboration phenomenon is often contradictory. Theories range from purely economic to those which entirely discount price and cost considerations. In evaluating the evidence, and critically examining the broadly differing theories of collaboration, it is argued here that technological collaboration:

(a) is, and will remain, a significant feature within industry and of industrial innovation. Technological collaboration has a long history. The present period of rapid and uncertain technological change increases the propensity of firms to collaborate, and this is facilitated by technology itself in the form of various electronic media.

(b) has seen its importance exaggerated, which detracts from consideration of its real value. Collaboration is no substitute for in-house technological efforts, and in comparison to internal R&D efforts its scale is very limited. Nevertheless, it can be valuable, assisting, for example, a greater pluralism in inputs to technology development and is particularly useful in integrating the specialist contributions of small firms.

(c) should be seen not only as a means of developing new products and processes, but very importantly as a way of improving technological capabilities. The enhancement of these capabilities is a target both of corporate strategies and public policies. The formulation of such strategies and policies has, however, to be shaped in the understanding that collaboration can have negative consequences. It can, for example, reduce innovation and be anti-competitive.

(d) is particularly valuable in encouraging firms to learn to do things differently. A major conclusion of this book, with relevance for the theory and practice of collaboration, is that collaboration provides an important and necessary stimulus to technological and organizational learning.

(e) has proved very difficult to manage successfully. The managers of few firms, outside of Japan, are comfortable thinking in terms of capabilities and learning. Such outcomes require the extension and development of forms of relationship between firms, and management styles which are long-term and based on high levels of trust. By focussing on short-term, individual product-related outputs, collaboration has inevitably failed to meet its often high expectations.

The structure of the book is as follows. Chapter 1 introduces the whole question of the significance of technology in industry, national differences and the challenge faced by firms. Chapter 2 introduces technological collaboration, looking at its extent, form and focus. Chapter 3 poses the question: why collaborate? It answers from within four perspectives: an innovation perspective, which is examined in greater detail in Chapters 5 and 6; a public policy perspective, studied in Chapter 7; a corporate perspective, which is further examined in Chapter 8; and an internationalization perspective, studied further in Chapter 9. Prior to these chapters, Chapter 4 reviews some of the theoretical and analytical approaches to collaboration. Chapter 10 begins a consideration of some particular issues in collaboration, in this case, collaboration in Japan. Chapter 11 examines technological collaboration in small firms, and Chapter 12 looks into the management of

collaboration. Chapter 13 examines the future implications of technological collaborations for firms and public policy.

Throughout the book six detailed case studies are presented to illustrate a
number of the most important issues in R&D collaboration, and also to put some
flesh on the bones of collaboration by describing how it comes about, the processes
involved in its undertaking, and the problems, particularly in its management.
These case studies provide much of the basis for Chapter 12.

The book does not address at any length the questions of links between firms
and research institutes. This is the focus of a forthcoming book on *Technology
Transfer in Europe* by the author and John Bessant, to be published by Routledge.

The writing of this book has been supported by the Economic and Social
Research Council's Designated Research Centre in Science, Technology and
Energy Policy at the Science Policy Research Unit. The case studies were funded
by the Centre for the Exploitation of Science and Technology (CEST), and
appeared in a much truncated form in Dodgson (1991a). Three of the case studies
(GEC, BT&D and Ricardo) were undertaken by Paul Simmonds, and one (Racal)
by Paul Quintas. My thanks are extended to them, to John Bessant, John Cheese,
Mari Sako and Nick von Tunzelmann, and to many more of my colleagues at SPRU
who kindly provided so much information and analysis. Particular thanks for their
advice and encouragement are gratefully offered to Margaret Sharp and Keith
Pavitt, and to Roy Rothwell who co-wrote much of Chapter 11, and whose work
on the innovation process infuses much of the book. The shortcomings in the book
are, of course, all mine. As ever, nothing could have been done without the
forbearance of my wife, Yo.

Mark Dodgson

Chapter 1

Technology in industry

This chapter analyses how important technology is to economic development and social well-being. It describes the marked differences in technological performance amongst industrial nations, and the primary contribution made to this by firms. It then describes some of the major challenges firms are facing in dealing with the complexities and difficulties of technological change.

THE ECONOMIC DIMENSION

In every industrialized nation the economic standard of living and the social quality of life depends crucially upon the way industry uses technology to enhance competitiveness. To this end nations and firms devote enormous resources to their technological activities. The OECD nations are currently spending over £170 billion annually on Research and Development (R&D), which employs many of the best-trained and creative people in each country. There are increasing numbers of scientists and engineers working in R&D and in total there are around one million R&D employees in Britain, France and Germany. There are roughly six million business establishments in the USA working in high technology industries, and scientists and engineers comprise over 4 per cent of the total USA workforce. This dedication to technological innovation is enormously important for economic prosperity and social welfare.

Technology and innovation are argued to be important factors influencing world economic growth cycles. Ever since the Russian economist Kondratiev noted in the 1920s the relationship between technological innovation and the roughly fifty-year economic cycles of prosperity, recession, depression and recovery, there has been a debate about the relationship between macro-economic activity and technological change. Economists such as Schumpeter have argued the importance of periods of intensive innovation in stimulating economic revival from recession. History has seen periods of 'clustering' of radical innovations. These major technological innovations, in conjunction with other circumstances such as war and famine, have been argued to be associated with economic cycles. Thus, steam power and textiles in the late 18th and early 19th centuries are associated with the first of what has become known as 'Kondratiev waves' of economic activity.

Railroad, iron, coal and construction innovations are associated with the second Kondratiev wave in the second half of the 19th century, and the period from then until after the Second World War is linked to the third wave development of the electrical power, automobile, chemical and steel industries. The post-war period has seen the fourth wave of continued innovation in automobiles, and developments in electronics and semiconductors, aerospace, pharmaceuticals, petrochemicals and synthetic and composite materials. Some argue that a fifth wave is going to be typified by continuing developments in communications and information technology, biotechnology, new materials, and computer-integrated manufacturing technologies.

Freeman and Perez (1988) describe the relationships between long waves of economic development and changes in techno-economic paradigm, which they refer to as

> a combination of interrelated product and process, technical, organisational and managerial innovations, embodying a quantum jump in potential productivity for all or most of the economy and opening up an unusually wide range of investment and profit opportunities.

(Freeman and Perez 1988:48)

They argue that recession and depression periods witness a mismatch between the possibilities of new technologies and organization of production and the existing social and institutional characteristics in industry. During a period of considerable adaptation and adjustment, these two areas are eventually integrated and are associated with economic recovery. For Freeman and Perez, therefore, it is not just periods of technological innovation that affect economic long waves, but also their combination with a number of economic and organizational changes.

Technology provides a means by which countries and firms compete internationally, and it underpins the remarkable re-alignment of national comparative advantage in the last thirty years. It is on the basis of its technological excellence that Japan has grown into an economic super-power. At the same time the declining international competitiveness of countries like Britain and the USA have been caused by an inability to remain technologically competitive. Between 1970 and 1986, for example, Japan's world share of high-technology manufactured products increased from 16 per cent to 32 per cent, while the USA's share declined from 51 per cent to 42 per cent, and the UK's from 8 per cent to 5 per cent (NSF 1989).

The level of technological activities in a country directly influences the wealth of nations. In a study of over 20 countries, Fagerberg (1987) found a statistically significant relationship between R&D and patenting performance and GDP per capita. He further argues that R&D and international patenting are significant determinants of differences between countries in export and productivity performance. Pavitt and Soete (1980) also show a positive statistical relationship between national technological activities and export performance.

THE SOCIAL DIMENSION

Technological development is intimately related to social and organizational development. Technological change derives from society and influences that society. Technology affects life at work and home. It can liberate us from repetitive, soul-destroying tasks, facilitate communications and travel, enhance the delivery of education and improve healthcare. At the same time hardly any technology is without negative implications. Technology has given us global warming, numerous local ecological disasters, machines of mass destruction, and the means to restrict and control civil liberties. It is also helping extend the wealth divide between developed and under-developed nations, between the technology-haves and have-nots.

The ways in which technology is developed and then used is of central importance for societies and economies. It is necessary to understand the ways in which technology is formulated and diffused through institutions and their organizational relationships in order to be able to influence it. Governments and industry and their agencies; individual firms and their managers, scientists and engineers; individuals as workers, consumers and voters all affect the way technology is created and shaped. To use the examples of the government and of individuals in the case of industrial innovation: governments play a vital role in shaping technology through their role as technology supplier (finance for the science-base), regulator (concerning intellectual property rights and technical standards, for example) and user (through procurement policy). Governments should, of course, be responsive to the pressures from their electorate, and individuals as voters can influence innovation in this manner. They can also be influential in other ways. The Japanese experience shows, for example, how important it can be to elicit feedback and new ideas from the workforce in order to enhance innovation. Consumer learning is a very important aspect of innovation, shown clearly in the differing fortunes of the new interactive service Minitel in France, which is now widely used by the public, and its British counterpart Prestel which only has a limited business function.

One current area where decisions of individuals as voters, workers and consumers are of increasing importance is that resulting from increasing environmental consciousness. Difficult decisions about, for example, the balance between continuing demand for automobiles and cheap electrical power and the dangers of CO_2 emissions and nuclear waste disposal need to be based on informed debate about technology as the cause of and solution to these problems. Knowledge is required at a national and international level of what the potentials of technology are, and what the political and corporate forces are which shape it. For example, what are the implications of international technological collaboration? On the positive side, can it offer global solutions to the global problems of over-population, pollution, and differential wealth and quality of life? Or, on the negative side, can it accelerate national disparities by excluding backward nations, and promote oligopoly and restrict competition in firms? Understanding the forces which shape technology is the basis of effective democratic decision making about it.

NATIONAL DIFFERENCES IN TECHNOLOGICAL PERFORMANCE

While many developed countries have benefited considerably from their creation and use of technology, other industrialized countries have fared less well. Before examining national differences it should be noted that there are considerable difficulties with all the measures of technological activities: R&D expenditure and employment, and patenting. These can at best be seen as piecemeal indicators, and are in no way comprehensive in their coverage (there are differences in what is recorded as R&D expenditure and who are R&D employees, and in the propensity of different firms and sectors to patent); nor can they be argued to cover all the technological activities of firms. Nevertheless, they are the best we have, and can be argued to be broadly indicative.

In its gross expenditure on R&D the USA predominates in the world, spending around one-half of total OECD gross expenditure. These figures, however, are limited as they do not reveal the expenditure dedicated to military aims, nor do they show trends. In 1989, 65 per cent of US government expenditure was military (compared with 45 per cent in the UK, 12 per cent in the Federal Republic of Germany, and 5 per cent in Japan) (Cabinet Office 1991). As for trends, the rate of growth in non-defence R&D between 1975 and 1988 increased by an average 10.9 per cent in Japan, nearly double that of the USA (5.8 per cent) and West Germany (5.5 per cent), and three times that of the UK (3.2 per cent) (NSF 1991). Gross R&D expenditures as a proportion of Gross National Product remained roughly the same in the USA and the UK between the early 1960s and late 1980s at around 2.5 per cent, while those of Japan and West Germany doubled during this period and actually overtook the USA's and UK's commitment in the late 1980s (NSF 1989).

Crucial in this change has been the commitment to technological activities of business firms, and the way that technology is managed and used. Business expenditure on R&D is the largest component of gross national expenditure in R&D in almost all developed nations, and its relative contribution is increasing compared to that of governments and higher education (see Table 1). Furthermore, since the Second World War business enterprise R&D expenditures have grown more rapidly than output growth in most OECD countries (Soete 1991).

As seen in Table 1.1, the majority of leading OECD nations' gross expenditure on R&D is undertaken in the business sector. However, there is considerable international variance in industry financing of R&D. Much higher proportions of national R&D efforts are funded by industry in Japan and Germany than in the UK and USA (Soete, 1991). The most revealing statistics in this respect relate to trends in industry-funded R&D. These are shown in Table 1.2.

The steady growth of industrial commitment to R&D seen in Japan, Germany and Sweden is in marked contrast to the relatively low and static figures in France, Italy and the UK. The contribution of US industry is at a medium level and does not show the growth enjoyed by the leading countries during the 1980s. However, individual US companies predominate in R&D spending. Table 1.3 indicates the extent of the commitment to technology by some companies, showing the five

Table 1.1   R&D expenditure by performing sector (percentage shares)

Country	1971	1980	Most recent year
		Enterprises	
France	56.2	60.4	61.3 (1990)
Germany	63.7	70.3	73.5 (1990)
Japan	58.4	59.9	67.9 (1988)
UK	61.3	61.8	66.6 (1988)
USA	65.7	69.3	70.7 (1990)
		Government sector	
France	26.9	23.1	24.9 (1988)
Germany	14.2	14.8	12.0 (1990)
Japan	12.4	11.8	8.8 (1988)
UK	25.2	22.1	14.4 (1988)
USA	16.2	12.1	11.1 (1990)
		Higher education	
France	15.6	15.4	14.8 (1988)
Germany	21.6	15.5	13.9 (1990)
Japan	27.6	25.5	19.0 (1988)
UK	11.0	13.3	15.1 (1988)
USA	14.9	14.5	15.5 (1990)

Source: OECD. STIID Data Bank (Quoted in Soete 1991)

Table 1.2   Trends in industry-funded R&D as a percentage of GDP in major OECD countries, 1967–88

	1967	1975	1985	1988
United States	1.01	1.01	1.35	1.38
Japan	0.83	1.12	1.84	1.95
Western Europe	0.79	0.83	1.07	1.17
France	0.61	0.69	0.94	0.96
FR Germany	0.94	1.12	1.58	1.78
Italy	0.35	0.47	0.58	0.54
Sweden	0.72	0.96	1.71	1.74
United Kingdom	1.00	0.80	0..96	1.06

Source: Patel and Pavitt (1991a)

largest R&D spending firms in Britain, Europe, the USA and Japan. It shows that US companies are the largest spenders on R&D in the world: nearly one-half of the world's top 200 R&D spenders are US companies (*Business Week* 25 October 1991).

Soete (1991) separates four groups of countries according to their relative R&D efforts. Using a measure of R&D Intensities (RDI): (Business expenditure on R&D/Gross Domestic Product × 100), 'technological leaders' are defined as having

Table 1.3   The largest corporate R&D spenders

	Current R&D expenditure (£ million)
Britain	
ICI	591
Shell Transport and Trading	473
Unilever	408
Glaxo	399
SmithKline Beecham	392
Europe	
Siemens (G)	2308
Philips Electronics (N)	1346
Alcatel Alsthom (F)	1249
Bayer (G)	949
Hoechst (G)	931
USA	
General Motors	2984
IBM	2745
Ford Motor	1987
AT&T	1359
Digital Equipment	901
Japan	
Hitachi	1682
Matsushita Electric Industrial	1353
Fujitsu	1171
Toshiba	1041
NTT	971

Source: Data derived from *Business Week* 25 October 1991 and Pari Patel.

Note: Exchange rate $1.79/£1

an RDI over 1.5; 'other high tech countries' as having an RDI of 1.0 to 1.5; 'middle tech countries', 0.5 to 1.0; and 'low tech countries', up to 0.5. Table 1.4 shows the countries in each group. The technological leaders are, with the exception of the USA, denoted by the increasing commitment from the private sector to R&D. In the USA, although both public and private sector R&D is increasing, the increase in the private sector is about half the increase in Japan or Germany (Soete 1991).

Using the measures of R&D and international patenting, Pavitt and Patel (1988) also describe a number of broad trends in national performance. Essentially, highest growth in these areas is seen in Japan, middle growth is seen in continental Western Europe in countries like Germany, Sweden and Switzerland, and lowest growth is in the USA, UK and Netherlands. They characterize national technological systems as 'dynamic' and 'myopic'. A number of inter-related reasons are suggested for determining this distinction. In myopic systems investment in technology is made on the basis of short-term financial criteria, with no understanding, as in dynamic systems, of the way technology 'includes the prospect of creating new market demands, and of accumulating over time knowledge and experience that open up

Table 1.4   National technological positions – OECD countries

Technological Leaders	Other High Tech Countries
Germany	Belgium
Japan	Finland
Sweden	France
Switzerland	Netherlands
USA	Norway
South Korea	Taiwan
	UK

Middle Tech Countries	Low Tech Countries
Austria	Argentina
Canada	Australia
Denmark	Greece
Ireland	Iceland
Italy	New Zealand
India	Portugal
Mexico	Spain
	Turkey
	Yugoslavia

Source: Soete (1991)

further technological applications and business opportunities in future' (Pavitt and Patel 1988: 51).

Other reasons include the adhesion of myopic systems to inflexible corporate structures, inappropriate technological and market management skills, and inadequate employee capability to learn from producing and using new technology.

It is therefore at the level of the firm that national differences in technological performance are determined. US predominance in technological expenditure is not translated into predominance in technological performance. The changes taking place in key industries and technologies, with growing Japanese and declining Western strength, occur because of what is going on in firms. As Porter (1990) says, there is no such thing as competitiveness of nations, only competitiveness of firms.

There are, of course, major differences in technological performance within countries. The USA retains its considerable strength in military and raw materials-related technology, while the Japanese, who have hardly any interests in these areas, are particularly strong in electronics, automobiles and mechanical engineering. Some European countries have particularly strong sectors, such as fine chemicals and pharmaceuticals in Britain and Switzerland; motor vehicles in Germany; and mechanical engineering in Sweden. The strength of these sectors is often allied to the technological activities of a few large firms (Pavitt and Patel 1988).

THE EXTENT OF THE TECHNOLOGICAL CHALLENGE FOR FIRMS

It is within firms that technology is developed and put to use, and the extent of the difficulties and uncertainties facing firms in their technological activities are enormous. Freeman and Perez (1988) in their discussion of the new 'techno-economic paradigm' of information technology describe its profound impact on existing firms and sectors and its ability to generate new ones. This new paradigm engenders the need for new organizational structures in order to adapt to and reflect the opportunities and threats provided by radical technological change. They consider that these new forms include greater 'networking' and technological cooperation (see Chapter 4).

The extent of these difficulties are compounded by individual companies' inability to control the development of technology and thereby restrain uncertainty. The largest corporate spender on R&D in the world, General Motors, which spent over $5 billion on R&D in 1990, accounted for only 3 per cent of USA expenditure.

Further difficulties arise due to the actual nature of technological knowledge. Technology is not just information which can easily be bought and sold, packaged, transferred and integrated. It is instead often knowledge which resides in individuals and in the routines and procedures of firms. It is commonly tacit knowledge which cannot be codified, and is highly specific to individual firms. Furthermore, it is cumulative over time (Pavitt 1991). This further differentiates firms' technological activities and affects the propensity to follow their previous paths of technological development. As Dosi (1988) says, 'what a firm can do in the future depends on what it has done in the past'. Technological development depends critically upon firms' ability to learn. These issues will be examined in more depth in Chapter 4, but it is worth noting that learning in firms is severely constrained by the tendencies of organizations towards introspection, and by the way strategies reflect existing ways of doing things, rather than novel approaches.

Innovation and technology has to be managed, and this is an increasingly complex task. The extent of scientific and technological knowledge is expanding rapidly, and the sources of this knowledge multiplying. There are estimated to be between 40,000 and 50,000 scientific journals currently being published worldwide. Industrial firms are increasing their investment in academic R&D. In the USA, industry more than doubled its expenditure in academic R&D between 1977 and 1987 in real terms (NSF 1987), and German industry has been increasing its external expenditure on R&D from around 3 to 4 per cent in the early 1970s to over 8 per cent in the mid and late 1980s (Hausler 1989). In Japan, where traditionally university/industry links have not been extensive, funds provided by industry for university research have increased very rapidly and are still increasing at around 20 per cent a year (*Nature* 1 November 1990), and the number of joint research projects increased from 56 in 1983 to 583 in 1988 (Hicks *et al.* 1992). Science and engineering articles co-authored by industry and academic contributors doubled as a proportion of all industry alone papers between 1973 to 1986, increasing from 13 per cent to 28 per cent (NSF 1989).

A very wide range of firms, including small ones, undertake R&D. It is estimated that there are 25,000 small and medium-sized enterprises (SMEs) undertaking R&D in (West) Germany (Kuhlmann and Kuntze 1991). There are over 1.5 million high-technology small and medium-sized establishments in the USA (NSF 1989). Small firms are important sources of innovation in particular industries and sectors (Acs and Audretch 1990).

The technology which is so critical to the economic and social well-being of nations is created and used within firms which are confronted by an immensely complex environment of potential suppliers and users of technology. The organizational challenges to deal with this complexity are similarly far-reaching. R&D managers, for example, frequently have to link the activities of their units with other functions within the firm, other R&D units in the firm and in other firms, both national and international, and they have to integrate new knowledge from universities, institutes of higher education and contract research organizations. The whole process of innovation and technological change has, as we shall see, become very much more complicated.

Chapter 2

Technological collaboration in industry

Technological collaboration is an activity firmly established within governments, firms and academic and research organizations. There has been an explosion of debate and analysis into the cause, nature and purpose of collaboration. The following chapters provide a review of the disparate literature on the role and potential of collaboration in industry, its extent, the ways governments and firms have promoted it, explanations for it, and analyses of its process and consequences. This chapter examines what collaboration is, the extent to which it is occurring, and its focuses.

WHAT IS COLLABORATION?

There is a plethora of definitions of collaboration – also known as 'alliances', 'cooperative agreements' and 'networks' – including a huge range of activities. They are formed by firms with other firms – suppliers, customers and, occasionally, competitors – and with higher education institutes and contract research organizations. Collaborations take place in the research, development, manufacturing and marketing functions, and take a wide variety of forms. Vertical collaboration occurs throughout the chain of production for particular products, from the provision of raw materials, through all the manufacture and assembly of parts, components and systems, to their distribution and servicing. Horizontal collaboration occurs between partners at the same level in the production process.

The technological collaborations discussed here are not simple contracts between organizations of once-and-for-all sales of a product, service or licence, but continuing arrangements where partners extend their expertise through sharing skills and personnel. The aims of these linkages may include increased knowledge of technological threats and opportunities, and improved capabilities in product development and efficiency in production. Collaboration can be conceptualized according to its aim, form or spatial nature. These categories are not mutually exclusive, and they may overlap.

The aims of technological collaboration include improvements in the innovation process, and the various technological objectives of corporate strategy and public policy. They encompass:

1 Improving the **Development Process**: collaboration in complementary areas arranged between different organizations (which may or may not be part of the same firm) sharing the development of new knowledge, products and processes. Such collaboration is found, for example, between automotive 'design houses' and assemblers (Graves 1988).

2 Enhancing efficiency in the **Production Chain**: collaboration in supplementary areas arranged vertically throughout the range of production activities. Examples are found between component manufacturers and car assemblers in the automobile industry (Lamming 1992), and within manufacturing in civil and military aircraft (e.g. Airbus and European Fighter Aircraft – the way that one firm provides the engine, one the wing, another the avionics, etc.). It may also manifest itself in subcontracting relationships between large and small firms, which is so important in Japanese industry (Economic Planning Agency 1990).

3 Merging previously discrete **Technologies and Disciplines**: such collaboration is found, for example, between mechanical and electronic engineering in the creation of 'mechatronics', in what Kodama (1987) calls technological 'fusion', and in the way biology, chemistry and physics have combined in the evolution of biotechnology.

4 Learning through **Information Exchange** about the potentials of collaboration and of particular partners: this may be facilitated in the first instance by the use of networks and databases, both public and private, designed to improve awareness of, and access to, technological knowledge, e.g. through patents and citations, so as to ascertain who knows what. An example of such exchange is provided by the European Community (EC) CORDIS database which includes information on Community-funded R&D results, projects and publications. Information is also exchanged about potential partners. The EC, for example, operates the BRE centre (Bureau de Rapprochement des Enterprises) and BCNet, providing profiles of potential collaborators.

5 **Corporate Strategies**: these may have the aim of using collaboration to improve innovation in the ways described above. They may be concerned with reducing the cost, risk and uncertainty of technological innovation. Strategies could be aimed at promoting collaboration between divisions and individual firms within a group to find synergies, or between large and small firms to merge the resource advantages of the former with the behavioural advantages of the latter. Collaboration may be designed to inform decisions about future mergers or acquisitions, and it may be used for competitor exclusion and strategic game playing (see Chapter 3). Internationalization of activities can be assisted by collaboration, as can corporate learning about new activities. For example, the joint General Motors/Toyota NUMMI manufacturing plant in the USA assisted Toyota's strategy of expanding in the USA, and General Motors' strategy of learning to improve manufacturing capability.

6 **Public Policies**: may be aimed at increasing collaboration in order to improve the comparative technological performance of national firms; to enhance the

efficiency of the national system of innovation by encouraging firms which ordinarily would not be involved in research to do so, and to enhance information flows between firms and between firms and the science base.

The forms of collaboration may include:

1 **Infrastructural** forms which are embedded in national technology and innovation systems, and are created especially to support that system. Thus the universities, government laboratories and independent contract research organizations which offer collective industrial research provide infrastructural forms of collaboration. Also included are public policy supported collaborative R&D schemes, for example, the Alvey Programme in Advanced Information Technology in Britain, and the ESPRIT Programme in the European Community. Also included would be Research Consortia of groups of companies sponsoring research of common interest, such as the Microelectronics and Computer Technology Corporation (MCC) in the USA. Whereas both forms of collaboration can be described as 'infrastructural' it is important to distinguish research which is undertaken within firms, and is coordinated, as in the cases of Alvey and ESPRIT, from joint research in joint research facilities, such as in MCC, as in the latter there is considerably more cooperation, and knowledge is jointly created and shared. Another form of collaboration in this sense are the publicly available information networks and databases (e.g. the EC's CORDIS system and Minitel in France).

2 **Contractual** forms, which may take the shape of a joint venture, formed by two or more partners as a separate company with shared equity investments. An example is Fuji–Xerox, a joint venture between Fuji Film and Rank Xerox. Management within joint ventures may be fairly autonomous but essentially they are controlled by the investing partners. It could be a partnership linking firms on the basis of continued commitment to shared objectives without equity sharing, commonly known as 'strategic alliances'. An example is the arrangement between IBM and Siemens to develop the next generation of 64 megabyte chip. R&D agreements occur between firms sharing R&D efforts in particular projects or programmes, and firms may agree to exchange technologies. R&D contracts can be a form of collaboration if both the contractor and subcontractor contribute technical know-how, or if the contract is flexible and alters with what is discovered. An example of the latter may occur if the contractor is continually learning from the subcontractor, and future research priorities are mutually agreed (see Chapter 5 for an example).

3 **Informal** forms, which are very important for the innovation process in the way they occur between the 'invisible college' of peers. Von Hippel (1988) describes the way that 'informal know-how trading' between peers occurs in a number of industries. 'Informal know-how trading is essentially a pattern of informal co-operative R&D. It involves routine and informal trading of proprietary information between engineers working at different firms – sometimes direct rivals' (Von Hippel 1988:6).

Informal collaboration can manifest itself within discussion groups and working parties; it is known to be important in a wide range of sectors and technologies, and has been argued to be a precursor to more formal partnerships (Kreiner and Schultz 1990).

The spatial nature of collaboration encompasses:

1   International collaborations: government or corporate decisions made to access specific international competences, or to circumvent national restrictions, or to benefit from international collaboration promotion (e.g. with the European Community).

2   Regional collaboration, which may involve coordinated attempts, often promoted by regional governments and agencies, to stimulate such collaboration and integration. This is particularly important in decentralized nations, such as Germany, where, for example, the Baden-Württemberg Land plays an important role, but there are increasing attempts at imitation by organizations such as the Welsh Development Agency and the 12 UK Regional Technology Centres. The regional agglomeration of firms working in similar industrial sectors and technologies and collaborating closely in the form of networks has been argued to have improved the innovative and competitive potential of regions in the USA, such as Silicon Valley and Route 128, and in Europe such as Emilia-Romagna in Italy, Jutland in Denmark, Southern France, and the Catalonia and Valencia regions of Spain.

3   Local collaboration, which involves coordinated attempts, often promoted by local and city councils, to stimulate such collaboration and integration. A typical manifestation of this may be found in the form of local science parks or, on a larger scale, the CRITT (Centres Regionaux d'Innovations et de Transfert de Technologies) in France, which are designed to encourage collaboration between public research bodies and industry and to help small and medium-sized firms innovate.

The full range and scope of collaboration is too broad to allow full consideration here. Instead a more restricted definition is used which includes any activity where *two or more partners contribute differential resources and know-how to agreed complementary aims*. In this definition may be included the following, both privately created and promoted by public policy:

(a) Collaborative research programmes or consortia.
(b) Joint ventures and strategic alliances.
(c) Shared R&D and production contracts.

Both vertical and horizontal linkages are included. Other linkages between firms, which are occasionally described as collaboration, are not covered here. Thus, direct investment, licensing, marketing agreements and computerized networks and data-banks which can improve technological capabilities are not included as they are essentially one-way transfers of know-how.

Although some collaborations involve many partners, the majority involve only a few. Levy and Samuels (1991) cite a report that shows 86 per cent of Japanese inter-firm research involves a single partner. Projects in the UK's Alvey Programme averaged 3.6 partners (Guy and Georghiou 1991). While many publicly-promoted *programmes* encompass large numbers of firms, individual *projects* within them tend to involve only a limited number.

Although collaboration occurs in many different forms and for very diverse motives, a number of generalizable assumptions underpin collaborations in the sense used here. First is the belief that collaboration can lead to *positive sum gains* in internal activities. That is, partners can together obtain mutual benefits which they could not achieve independently. Benefits may include increased scale and scope of activities, reduced costs, greater speed or improved ability to deal with complexity. Second, collaboration is believed to assist with *environmental uncertainty*. Increasingly sophisticated and demanding consumers, the growing competition in and internationalization of markets, and rapidly changing and disruptive technologies place pressures on nations and firms to exist with, and attempt to control, the uncertainties confronting them. This is more easily achieved in collaboration than in isolation. Third, collaboration is believed to offer *flexibility* in comparison to its alternatives. So, for example, it may be an alternative to direct foreign investment and to mergers and acquisitions which are less easily amended once entered into. It can facilitate the transfer of know-how between organizations more effectively than can purely market transactions. These assumptions will be examined in greater depth in subsequent sections.

THE EXTENT OF COLLABORATION

Technological collaboration is nothing new. It was a feature of industrial activity identified by Alfred Marshall last century. Alic (1990) refers to the first recorded attempt to put together an R&D consortium: that of Josiah Wedgewood with a group of Staffordshire potters in 1775. International collaboration also has a long record. ICI and Du Pont signed a technology sharing agreement in 1929, and Bosch and Nippondenso formally began sharing their technology in 1953. Freeman (1991) points out that there has been a history of formal R&D collaboration from the 1920s and that these have provided the focus of academic analysis and debate in the 1950s, 60s and 70s. The growth of formal collaborative research organizations began in earnest in the Research Associations created in Britain following the First World War. This model was transferred to Japan where there has been a history of inter-firm cooperation and linkages in the pre-Second World War *Zaibatsu* and post-war *Keiretsu*. The success of cooperative research in Japan led to attempted replication in the USA and in Europe (a classic case of this was the USA's and EC's action in forming collaborative research programmes in response to Japan's Fifth Generation Computer Systems Programme (Arnold and Guy 1986)).

What is new is the extent of collaboration which, as we shall see below, increased throughout the 1980s, the degree to which technological considerations stimulate and facilitate collaboration, the changing strategies of firms which place less emphasis on total integration of activities, the active promotion of collaboration by governments, and an increasingly international perspective within firms and nations.

Establishing the extent of collaboration is hampered by the paucity of sound information available. It is known that some governments and firms actively promote collaboration. Japanese high technology industry has a plethora of collaborative R&D programmes and organizations (Levy and Samuels 1991). In the USA, over 170 consortia have been formed since the 1984 Cooperative Research Act was introduced to shield them from anti-trust legislation (Werner and Bremer 1991). The European Commission has actively promoted collaboration through its Framework Programme, some of the major programmes within which are shown in Table 2.1. The Department of Trade and Industry (DTI) in the UK promotes over 100 industrial 'clubs' (Rothwell, Dodgson and Lowe 1989).

Some individual firms are known to collaborate widely, even when they have worldwide reputations for their in-house excellence in R&D. For example, in 1991 ICI had 30 international collaborations. IBM appears increasingly to use partner-

Table 2.1   1990–4 EC Framework Programme

	Budget £m
Enabling technologies	
Information and communications technologies	
Information technologies	946
Communications technologies	342
Development of telematics	266
Industrial and materials technologies	
Industrial and materials technologies	524
Measuring and testing	98
Management of natural resources	
Environment	
Environment	290
Marine sciences and technologies	73
Life sciences and technologies	
Biotechnology	115
Agricultural and agro-industrial research	233
Biomedical and health research	93
Life sciences and technologies for developing countries	78
Energy	
Non-nuclear energies	110
Nuclear fission safety	139
Controlled nuclear fusion	320
Management of intellectual resources	
Human capital and mobility	363
Total	£3990 million

Source: DTI (1991)

ships in order to retain its international technological competitiveness, working with, amongst others; Apple, Motorola, Groupe Bull, Siemens, NTT and Ricoh. Olivetti claims to have no less than 229 technological alliances (*Financial Times* 24 March 1992). Some Japanese companies like NEC and Toshiba have used collaboration extensively in their diversification and internationalization strategies.

Details on trends in collaboration in particular sectors rely on studies using questionnaire and case study approaches and are therefore sometimes highly selective. Aggregate information has depended mainly on a number of databases constructed of *reported* collaborations in the financial and technical press. These inevitably are restricted in the range of sources used, and do not include what some managers believe to be the large number of unreported and secret collaborations. Furthermore, they tend to report *numbers* of collaborations rather than their size and hence significance. These databases do not use common definitions of collaboration, and their findings have occasionally been contradictory. Nevertheless, the databases on collaboration do provide a broad indication of developing general trends and their focus. And as Chesnais (1988) says

> Flexibility in definition and in the range of agreements covered may, in fact, reflect one aspect of reality. The responses firms can give, through co-operation with other firms and organizations, to the need for a broadened base of scientific and/or technological resources and ensured access to important markets, are extremely varied. They will very often differ from one industry supply structure and/or one major type of technology to another.
>
> (Chesnais 1988: 55)

Chesnais (ibid.) quotes six studies of inter-firm agreements which used a variety of research methodologies. A general picture emerges of increased numbers of new collaborations being formed in the early and mid-1980s. This increase is also accepted to be the case by Mowery (1988), Contractor and Lorange (1988) and Mytelka (1991) who believe this trend may continue into the 1990s. For data in the late 1980s we are dependent on the MERIT–CATI database of over 7000 agreements reported in the press and specialist journals (Hagedoorn and Schakenraad 1990, 1992). Figure 2.1 shows the increase revealed by the other studies in the early and mid-1980s continuing in the late 1980s with, perhaps, the rate of increase slowing. The definition of collaboration used in the construction of this database includes some activities which, using the definition adopted here, might not be construed as strict collaborations. For example, one-directional technology flows (including licences) are included. Furthermore, it is not possible to assess the quality of the database in its coverage and construction and there are some anomalies in its reporting. Nevertheless, the MERIT–CATI database does provide some of the most comprehensive and up-to-date information on collaborations currently available.

One of the reasons why the rate of increase in numbers of collaborations being formed may be slowing (as is argued by Hagedoorn and Schakenraad) will be examined in subsequent sections of the book, i.e. the sheer management effort needed in order to make them work. Other reasons suggested include an end to pre-1992 repositioning amongst European multinationals, restricted numbers of available partners following the 1980s acquisitions and mergers boom, and decreasing uncertainty resulting from maturing technologies. Collaboration remains, however, an important element in the strategies of many firms and governments, and these arrangements are very often intricate and complex, as shown in Figure 2.2.

THE FOCUS OF COLLABORATION

Studies of collaborations across industry show the high number of them devoted to technological issues. Table 2.2 reports the findings of nine studies into cooperative agreements. It shows that many have a technological focus and that, very roughly, around one-quarter to one-third of them involve joint R&D. A high proportion of the agreements are shown to be in information and communications technology.

Mowery (1988) suggests that technology is increasingly the focus of collaborations and that technological collaboration is appearing in a wider range of industrial sectors and firms. Harrigan (1986) sees collaboration as a feature of high tech-

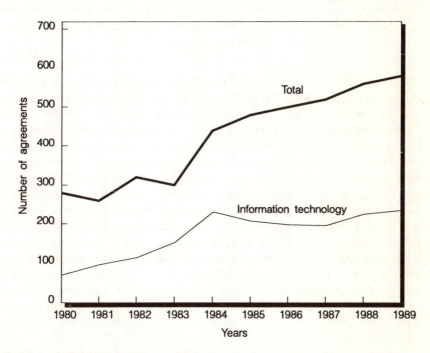

Figure 2.1   Growth of newly established technology cooperation agreements in general and in information technology
Source: Hagedoorn and Schakenraad 1992

Table 2.2 Studies of cooperative agreements

Study	Sample	Industry/technology		Technology focus	Joint R&D
Chesnais (1988)	Review of 6 databases on cooperation			'Over half' of cases	
Baughn and Osborn (1990)	270 US–Japanese cooperations	Motor vehicles Computers Chemicals Telecoms Banking	10.7% 10.3% 9.2% 7.0% 6.2%	'Emphasis' on technology	21%
Hegert and Morris (1988) INSEAD	839 international collaborations	Motor vehicles Aerospace Telecommunications Computers Other electrical	23.7% 19.0% 17.2% 14.0% 13.0%	'largest number' involve joint developments	37.7%
Harrigan (1986)	895 US strategic alliances	Petrochemicals Electronic components Financial services Communications services Computers	14.2% 12.1% 8.0% 7.2% 4.9%	One-quarter in electronics and communications	
Hagedoorn and Schakenraad (1990) MERIT	Over 7000 cooperative agreements			Over 60% in IT, biotech., new materials	27.5%
Mytelka (1991) LAREA/CEREM	1086 inter-firm agreements	Biotechnology, IT, civil aeronautics, automobiles			29% focus on knowledge production and sharing
Hakansson (1989)	123 small and medium sized Swedish firms	Manufacturing industry		Over one-half took more than 65% of development work collaboratively	
Daniels and Magill (1991)	187 US firms	High technology		12% had international joint ventures	
Kleinknecht and Reijnen (1991)	1929 firms	Representative of Dutch industry		No correlation between collaboration and: R&D intensity R&D intensive industry	25.9% (domestic) 10.4% (international)

SEMICONDUCTOR MERGERS AND ALLIANCES

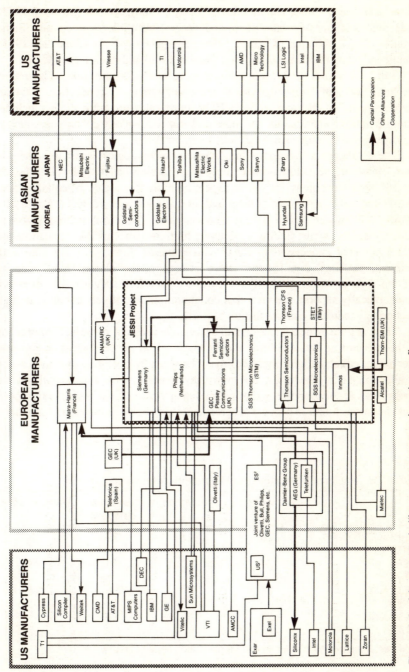

Figure 2.2 Mergers and alliances among semiconductor firms

nology industry, and of the development and early use of new technologies. Studies of the following individual industries and technologies show a high level of collaboration: information technology (Freeman 1991); biotechnology (Pisano, Shan and Teece 1988); automobiles (Womack 1988); aircraft (Mowery 1987); telecommunications (Pisano, Russo and Teece 1988); integrated circuits (Steinmuller 1988); robotics (Klepper 1988); computer systems (Saxenian 1991), semiconductors (Hobday 1991); food (Senker 1986) and steel (Lynn 1988).

Table 2.3 shows the broadly-defined focus of four studies into collaboration broken down into: knowledge (R&D, technology transfer); production; and marketing (including sales and servicing). The fourth category multiple/other includes multi-purpose collaborations and those undertaken for additional purposes such as 'globalization'. Sample sizes are included in brackets.

These data again suggest the importance of technological issues in collaboration. However, the picture is not entirely clear. Daniels and Magill's (1991) study of 187 high technology US firms, expecting a high incidence of collaboration, found that only 12 per cent of them had international joint ventures. And Kleinknecht and Reijnen's (1991) study of a sample of 1,929 firms representative of Dutch industry, whilst finding a relatively high incidence of R&D collaboration (25.9 per cent cooperated in R&D with domestic firms, 10.4 per cent with international firms), found no correlation between collaboration and R&D intensity of the firm or industry. Available evidence therefore suggests, but does not unequivocally support, the view that collaboration is relatively common in high technology industry, and is a feature of R&D.

There is little evidence to ascertain the relative levels of vertical collaborations – those occurring between firms at different levels of the production chain – and horizontal collaborations: those occurring between firms at the same level of the production chain. The little evidence we have tends to show a marked difference in the focus of collaboration in Europe and Japan. A survey of 839 mainly European collaborations found 15 per cent of agreements were between buyers and suppliers,

Table 2.3 Focus of international collaborations (percentage)

	Knowledge	Production	Marketing database	Multiple/Other
LAREA/CEREM (1086)	29	29	27	40
FOR (974)	34	23	14	28
INSEAD (839)	37	23	8	24
Takeuchi (4709)	20	26	30	*

Sources: Chesnais (1988); Hegert and Morris (1988); Mytelka (1991).

while 71 per cent were horizontal agreements between rivals. 14 per cent were devised for new market entry (Hegert and Morris 1988). In comparison, Levy and Samuels (1991) argue that four-fifths of inter-firm research collaboration in Japan is vertical, involving firms operating at different phases of the production process, and only one-fifth is horizontal between competitors.

The European figures may disguise the importance of vertical collaboration in particular industries. The importance of technological inputs of suppliers and users into firms' innovative activities are well known (Von Hippel 1988). And studies of a number of industries such as automobiles and consumer electronics show extensive vertical collaborations between assemblers and component manufacturers (Lamming 1987; Sako 1992). This anomaly points to one of the major shortcomings in counting collaborations recorded in the press, which is much more likely to be interested in activities between competitors, than in less newsworthy (and perhaps more historical) links between non-competing firms.

There is debate concerning the type of research most appropriate to collaboration. On the one hand, basic research is argued to be a public good, and thus a legitimate focus for public policy support, and furthermore as it is far-from-market it is less likely to cause problems for competing firms. However, applied near-market research is what firms are most interested in. Evidence on the relative focus of each type of research in collaboration is sparse. However, Levy and Samuels (1991) cite a Japanese government survey that found 14 per cent of inter-firm collaboration was directed at basic research, one-third involved applied research, and more than one-half was devoted to product development. They contend that although the situation is changing 'Applied research is typical of purely private agreements, whereas government-backed initiatives tend to be more oriented towards basic research' (Levy and Samuels 1991: 121).

Broad differences are suggested in the focus of collaboration between industries and technologies. Mowery (1988) highlights these differences with studies showing that in telecommunications, integrated circuits, commercial aircraft and robotics the focus of collaboration is product development; in automobiles and steel the focus is the production process; and in biotechnology and pharmaceuticals it is marketing and distribution. Other studies argue the way that the focus of collaboration changes over time. Hamilton, Vila and Dibner's (1990) study of biotechnology argues that the focus of collaboration changed over time in biotechnology firms from scientific to commercial activities. Kogut (1988), James (1989) and Mody (1990) contend that the focus of collaborations alters with product life cycles.

The question of whether collaboration is a strategic or a tactical concern, i.e. whether it focusses on issues important for the long-term development of the company or not, is one confused by the different definitions of collaboration used in different studies. There is a literature emphasizing the essentially strategic nature of collaboration, and another describing it as purely tactical, but often there is a mismatch in what is actually being examined. The consequences of a collaboration designed quickly to access a specific piece of technology at minimum cost are very different from those of, say, Du Pont, which uses collaboration to assist its diversification into electronics (see Chapter 8).

It is suggested that as technology is a strategic concern, collaboration in it is *de facto* strategic. However, Harrigan (1986), in one of the most systematic and comprehensive studies of joint ventures – the most formal contractual form of collaboration – argues:

> The closer activities are to a firm's strategic core – the higher their importance to the firm's survival – the less likely the firm is to rely on the research success of ventures or other arrangements with outsiders. The closer the research area is to a partner's strategic core, the more concerned the partner is with losing control of knowledge pertaining to those technological applications. In areas of high strategic importance, partners will make deals – licensing, cross-marketing, or other arrangements that they can tightly control – but they are less likely to create joint ventures.
>
> (Harrigan 1986: 148)

Other studies, however, argue that technological collaborations tend to occur in strategically important areas for firms, and for strategically important reasons. Baughn and Osborn's (1990) study of 112 USA/Japanese alliances argued that more than one-half of the agreements addressed core business areas of the two partners. Table 2.4 shows the data. The R&D expenditure as a percentage of the sales figure was obtained for the companies' core business area, and then the figure for the focus of the agreement was ascertained to be higher or lower. The evaluators of the UK's Alvey programme found that many of the industrial participants collaborated in their strategic technologies (Guy and Georghiou 1991). Hagedoorn and Schakenraad (1990) identified the purpose of the collaboration in their study and found that rather than focussing on short-term issues such as cost reduction, they tended to be more concerned with strategic issues such as long-term positioning.

Table 2.4   Technological intensity of alliance product relative to firm's core industry

| | | Japanese firm | | | |
		Lower	Same	Higher	
US firm	Lower	5	7	6	18 (16%)
	Same	4	55	19	78 (70%)
	Higher	1	6	9	16 (14%)
		10 (9%)	68 (61%)	34 (30%)	112 (100%)

Source: Baughn and Osborn (1990)

Some of the problems of this conflicting evidence can be overcome if account is taken of two issues. The first is the distinction between collaboration to build capabilities within firms, and those that focus purely on limited project-based objectives. Harrigan is surely right to say that firms will not cooperate in core areas of technology where product applications are 'near market' (unless markets can be clearly separated, for example, geographically). However, if the focus of the collaboration is 'pre-competitive', or is concerned with building capabilities which will in the future be used in the development of new products, then this form of strategic collaboration in core areas appears to make more sense. Furthermore, the aims of joint development of individual products in which a market already exists are correctly described as 'tactical'. But if the company, through one or more tactical collaborations, manages to develop its capability for future, and unforeseen, product developments, then these can be described as 'strategic'.

The second issue concerns the way the focus of collaboration changes over time. In one of the most sophisticated studies of the focus and form of collaboration, Cairnarca, Colombo and Mariotti (1992) argue that these vary along with industrial and technological development. Based on a study of over 2000 agreements in information technology between 1980 and 1986, they develop a technology life cycle model and relate collaborative activity to it. These various stages, and their consequences for collaboration, can in a simplified form be summarized as:

(a) Introduction: the first introduction of early, pioneering applications on to new markets, when there is considerable technological uncertainty. The propensity towards collaboration is very high, and the agreements focus on R&D, technical standards and 'technology watching' to see how things are developing. At this stage many of the collaborations involve equity agreements.
(b) Early Development: when market growth is very rapid, and technological opportunities are still very high. The propensity to collaborate is at its highest, and this is manifested in large numbers of non-equity investments to allow firms rapid access to specialized know-how in partners. Agreements focus on R&D, standards and joint development.
(c) Full Development: in which technological uncertainty is much reduced, and markets are expanding less rapidly. The number of agreements in relation to the size of the market declines, firms pursue strategies of internalizing know-how, and there is a contraction of non-equity forms of collaboration.
(d) Maturity: which occurs as market expansion slows right down, and technological efforts focus on using the knowledge accumulated in earlier periods. Non-equity collaborative agreements increase as firms attempt to revitalize technology and to exploit existing technology in peripheral markets.
(e) Decline: which is marked by market contraction and exhausted technological development. Firms are rationalizing and concentrating their efforts. Agreements between firms decline, and equity-based arrangements predominate in a wide range of commercial and manufacturing agreements.

The authors suggest that their evidence supports this life cycle model, although they admit that its empirical testing poses serious methodological problems. They nevertheless provide one of the most plausible analyses of the changes purposes and forms of technological collaboration.

CONCLUSIONS

In the examination of technological collaboration it is necessary to distinguish between vertical and horizontal linkages. It is helpful to conceptualize collaboration according to its aim, form and spatial nature. The aims of collaboration may include: improving the development process and production efficiencies, merging previously discrete technologies, information exchange, and there can be corporate and public policy aims. The forms of collaboration may be infrastructural, contractual and informal. It can be international, regional or local. The definition of collaboration used here includes activities where two or more partners contribute differential resources and know-how to agreed complementary aims. Collaboration provides positive sum gains, assists with environmental uncertainty and is flexible compared with its alternatives.

Despite a large and growing literature on technological collaboration, there still remains a somewhat hazy picture concerning trends in the numbers of collaborations, their focus and form, and their technological basis and strategic nature. However, it is argued here that collaboration grew in importance during the 1980s, and that technology is a central issue in this growth. The confusion in the contradictory evidence concerning the focus of collaboration is overcome once account is taken of the way collaboration assists the development of firm capabilities – a strategic issue – and the way the focus of collaboration alters with industrial and technological development. The reasons why collaboration has increased in importance can be explained by technological considerations, changing firm strategies, internationalization and public policy promotion. More light is thrown on these issues in the next chapter which looks at the reasons for the promotion of collaboration.

Chapter 3

Why collaborate?

If there is one thing which all the disparate literature on collaboration agrees upon it is that collaboration is a very difficult thing to make work. The management problems of collaboration are extensive. Furthermore, shared activities imply shared future income streams, and can compromise proprietary know-how. So why do firms collaborate? Collaboration can promote cartelization and oligopoly and raise entry barriers to new entrants. So why is collaboration so actively promoted by governments? Most studies of collaboration include a list of reasons and motives why firms collaborate (Harrigan 1986; Contractor and Lorange 1988; Mowery 1988). Reasons can be suggested for technological collaboration from within an 'innovation', 'corporate', 'public policy' and 'internationalization' perspective. Again, these categories are not mutually exclusive and there is considerable overlap. This chapter introduces the wide range of reasons for technological collaboration. Some of the issues raised here will be discussed in greater detail in subsequent chapters.

AN INNOVATION PERSPECTIVE

A number of issues raised within this perspective are briefly discussed below.

Linkages in the innovation process

Successful innovation requires innovating firms to have an external orientation. A common feature of the studies undertaken from the 1950s into successful innovation is the extent of external inputs – from customers, suppliers and academia – into internal innovative activities (Carter and Williams 1957; Rothwell *et al.* 1974; Maidique and Zirger 1985). In Gibbons and Johnston's (1974) study it was argued that external information inputs are as, if not more, important to innovative activities. Successful innovation depends on effective interactions between organizations (Lundvall 1988). Partnerships with suppliers can provide privileged access to state-of-the-art components. Strong links with important customers facilitate effective feedback on market requirements and product performance. Collaboration with other firms, perhaps even with competitors, and with university, government and private research laboratories, can extend a firm's options in innovation. Such linkages are nothing new. They may, however, be extending and intensifying, and may involve increasing use of information technology in cementing them.

Rothwell (1992) argues that we are heading towards a *Fifth Generation Innovation Process*. Some of the bare details of the various generations of innovation process are shown in Figure 3.1. The first two generations saw innovation as a sequential process, driven by technology or pulled by market demand. The third generation began to understand the way that technological possibilities and market requirements are coupled in a much more interactive way. The fourth generation of innovation process began to be seen in the second half of the 1980s, and was based to a significant extent on the Japanese experience. It involved parallel development of simultaneous R&D and prototyping and manufacturing, and the closer integration of firms. These linkages were both vertical – with suppliers and

---

First Generation:
Technology Push: Simple linear sequential process. Emphasis on R&D. The market is a receptacle for the fruits of R&D.

Second Generation:
Need Pull: Simple linear sequential process. Emphasis on marketing. The market is the source of ideas for directing R&D. R&D has a reactive role.

Third Generation:
Coupling Model: Sequential, but with feedback loops. Push or pull or push/pull combinations. R&D and marketing more in balance. Emphasis on integration at the R&D/marketing interface.

Fourth Generation:
Integrated Model: Parallel development with integrated development teams. Strong upsteam supplier linkages. Close coupling with leading edge customers. Emphasis on integration between R&D and manufacturing (design for makeability). Horizontal collaboration (joint ventures etc.)

Fifth Generation:
Systems Integration and Networking Model (SIN):
Fully integrated parallel development. Use of expert systems and simulation modelling in R&D. Strong linkages with leading edge customers ('customer focus' at the forefront of strategy). Strategic integration with primary suppliers including co-development of new products and linked CAD systems. Horizontal linkages: joint ventures; collaborative research groupings; collaborative marketing arrangements, etc. Emphasis on corporate flexibility and speed of development (time-based strategy). Increased focus on quality and other non-price factors.

Figure 3.1   The Fifth Generation Innovation Process
Source: Rothwell (1992).

customers in complementary activities, and horizontal – with competitors and firms in other industries in similar activities.

The major features of the Fifth Generation Innovation Process, which Rothwell calls the SIN model (Systems Integration and Networking), are that in order to produce new products flexibly, quickly and to high quality, strategic linkages between firms are needed, and these are often assisted by the use of expert systems and integrated Computer Aided Design/Computer Aided Manufacturing systems between suppliers and users. Mowery argues that inter-firm cooperation is considerably assisted by reducing costs of information transmission, storage and analysis.

> The exchange of technical, testing, and other data between development teams and the use of computer-aided design and manufacturing technologies in both development and production have made easier the 'spinning off' to other foreign or domestic firms of numerous tasks in the design and manufacture of complex products.
>
> (Mowery 1988: 16)

High costs of technological development

New technologies are extremely expensive to develop. A new drug can cost around $200 million to develop. A new semiconductor wafer fabrication plant can cost $500 million. Collaboration can help share these high costs, although returns from them will, of course, also be shared. Cooperation can reduce the unnecessary duplication of R&D efforts (Harrigan 1986).

Technological complexity and novelty

Many new technologies involve the conflation of previously discrete areas of knowledge. The development of 'mechatronics', for example, required the marriage of mechanical and electronic engineering, and attempts are being made to converge the new information and communications technologies. Currently research into high temperature superconductors involves the collaboration of materials scientists, physicists, chemists and electrical engineers. Few firms have the *breadth* of knowledge in the wide range of technologies in which convergence is possible, and so collaboration is a means of gaining access to these skills. Although some large companies may possess in-house know-how across the range of technologies which provide the focus of a collaboration, cooperation may be chosen by the large firm as a means of accessing *depth* of knowledge. Collaboration may be a method by which a potential product's *scope*, both in the sense of improved technical capacities and market applications, is expanded beyond what was possible through entirely internal development. (An example of this is provided in Chapter 6.)

Many new technologies are commonly systemic in nature, and while individual firms may have competences in some parts of the system, they could often require inputs from other firms. Saxenian (1991) offers the following example:

> A computer system today consists of the central processing unit (CPU) which includes a microprocessor and logic chips, the operating system and applications software, information storage products (disk drives and memory chips), ways of putting in and getting out information (input-output devices), power supplies, and communications devices or networks to link computers together. Although customers seek to increase performance along each of these dimensions, it is virtually impossible for one firm to produce all these components, let alone stay at the forefront of each of these diverse and fast changing technologies.
>
> (Saxenian 1991: 425)

Some argue the increased scientific content of technologies (Narin and Noma 1985), and this will increase the likelihood of collaboration for those firms without basic scientific competences. Teece (1988) refers to the many reasons why firms undertake R&D in-house, but he also discusses the way in which collaboration

> may represent an imperative in instances where the firm contemplating conducting research lacks the desired skills and is unable to acquire them in the labour market. This might actually characterize even research-intensive firms when a shift in the technological paradigm renders the existing skill base of the enterprise obsolete or irrelevant.
>
> (Teece 1988: 278)

The pervasiveness of information technology

Information technology pervades virtually every activity in industry, and the degree of commonality it engenders in the activities and methods of firms facilitates collaboration. Freeman (1991) argues that IT has

> found applications in every manufacturing and service sector, often changing profoundly both products and processes, but it also affects every function within each firm: design (CAD); manufacture (robotics, instrumentation, FMS, control systems, CIM, etc.); marketing (computer-based inventory and distribution systems) accounts and administration (management information systems, etc). Finally, it affects, through its convergence with the telecommunications system, the network of communications within the firm and its supplier networks, technology networks, customer networks, etc. In this last area it provides entirely new possibilities for rapid interchange of information, data, drawings, advice, specifications, and so on between geographically dispersed sites via fax, VANs (Value-Added Networks), electronic mail, teleconferencing, distance learning, etc. It is hardly surprising, therefore, that taking into account both the pervasiveness of IT and its systemic characteristics, most of the new develop-

ments in networking in the 1980s have been associated in one way or another with the diffusion of this technology.

(Freeman 1991: 509)

Information Technology is, of course, an international phenomenon. Imai (1990), for example, describes the way that strategic alliances between US and Japanese firms have substantially increased, and that this points to the

best institutional arrangement for coping with the emerging techno-economic system: a cross-border network enables firms to manage the complex process entailing an amalgam of cross-border, interfirm interactions. In particular, it works within the generic and complex industries distributed across information technologies.

(Imai 1990: 197)

Technological risks

Because of the high cost and complexity of much new technology development, it is a highly risky undertaking. Collaboration shares the risk of development. This has been seen in biotechnology, and it also occurs in technologies which are in the process of being widely diffused, such as information technology. There is considerable uncertainty as to the most appropriate configurations of the technology, and the markets in which it is to be used. This uncertainty may be overcome by collaboration (Freeman 1991).

The nature of technological knowledge

Much technological knowledge is tacit (Polanyi 1962) – that is, it is difficult to codify in the form of blue-prints, etc. – and firm specific (Pavitt 1988) and is, therefore, hard to transfer easily or quickly. Collaboration potentially provides a mechanism whereby close linkages between different organizations allow sympathetic systems, procedures and vocabulary to develop which may encourage effective transfer of knowledge. It allows partners to 'unbundle' discrete technological assets for transfer (Mowery, 1988). Technological knowledge is also difficult to price, and collaboration provides a means of exchange without necessarily resorting to prices.

Speed to market

Collaboration is believed by some to improve the speed of technological development. The Alvey Programme in the UK, for example, has been shown to decrease development times (Guy and Georghiou 1991). However, there is generally little evidence to suggest that this is universal, and the extent of the management problems involved in collaborations (discussed in Chapter 12) might tend to point to the lengthening of development times in collaboration. This points to one of the many trade-offs managers have to make in using collaborations. Involvement in them might lengthen development times, yet absence from them may lead to technological backwardness.

Creating technical standards

With many new technical standards being created, both domestic and international, firms may feel their cases for the adoption of particular technical standards may be improved by their promotion by a number of firms, rather than singly. This is particularly apparent within the EC. Collaboration provides an effective mechanism for the joint creation and promotion of standards (an example is provided in Chapter 6).

Apart from all these reasons suggesting why collaboration may influence innovation positively, there is an argument that collaboration is actually anti-innovation (Macdonald 1992). Macdonald argues that the legal stipulations on collaborations may restrict the free and easy informal collaboration which takes place between peers in a wide range of firms and is necessary for innovative efficiency. This may be the case in some collaborations, but as will be seen in subsequent chapters, there are a number of advantages in keeping the legal nature of collaborations as limited as possible.

A CORPORATE PERSPECTIVE

Changing strategies

The relationship between company strategies and technological collaboration is one which has excited a great deal of attention. Some argue that collaboration is critical for continuing competitiveness. Ohmae (1990), for example, in his argument about growing globalization (discussed below), suggests that it 'mandates alliances, makes them absolutely essential to strategy'. (1990: 114). Others develop a similar line: 'To find the right international partners has become a central strategic issue for most firms, an issue which is as important as the level and direction of spending on research and development' (van Tulder and Junne 1988: 250).

This is exaggerated. In-house R&D is, and will remain, the base for firms' technological accumulation (this is discussed in Chapter 13). Collaboration can provide a useful supplement to this base, and to this end firms are increasingly developing strategies for acquiring R&D knowledge externally (Friar and Horwitch 1986; Teece and Pisano 1987). '...the growth of world competition in technology-related areas makes (the) strategy of sole reliance on internally financed and internally conducted R&D insufficient and perhaps suicidal' (Link and Tassey 1987: 10).

Technological diversification

Studies of large R&D-undertaking companies have found a high level of techno-logical diversification during the 1980s (Pavitt *et al.* 1989; Granstrand and Oskarsson 1991). One of the most detailed and systematic studies of technological diversification examined 13 of Sweden's largest R&D spenders (Granstrand *et al.* 1990). It defined technological diversification as a firm's expansion of its com-petence into a broader range of technological areas. Using a measure based on the employment of qualified engineers in different disciplines, it found that 11 of these companies actively moved into new areas during the 1980s. The reasons for this diversification were growing international competition demanding increased pro-duct performance and/or decreased production costs, and rapid developments in technologies such as micro-electronics, biotechnology, new materials, and produc-tion technologies. Technological diversification was associated with increased R&D spending and was explained in the way that

> it is probably more difficult and expensive to combine R&D activities from different generic technology areas than to manage the same technology, another [reason] is the temporary fusion and substitution of new technologies [which] give temporary increases in investments. Also learning and scale effects prob-ably occur in the old set of technologies within the firm, which the corporation cannot benefit from when the technology is new.
>
> (Granstrand *et al.* 1990: 17)

Collaboration is an important weapon in the strategic armoury of firms attempting technological diversification. It can, in combination with in-house efforts, help overcome many of the problems of combining new technologies, and as will be argued below and in subsequent chapters, can speed up the learning process in firms. The importance of collaboration in technological diversification is seen in the activities of Japanese companies, where joint research programmes amongst firms in different industries are on the rise, and are explained by attempts to link different stocks of knowledge to generate new technologies (Aoki 1988).

Technological and organizational learning

As Grandstand *et al.* (1990) suggest, firms have difficulty in managing to integrate competences or technological knowledge in novel areas. Furthermore, firms tend to be organizationally conservative and stick to what they know best (see Chapter 4). In order to stimulate and facilitate learning to deal with new technologies and other changes affecting their operating environment, firms may pursue strategies of encouraging learning through collaboration (Dodgson 1991b, 1991c).

Competitor exclusion

Collaboration is a strategic tool which may be used to block competition, either by raising the scale of resources devoted to a project to deter other firms from attempting to compete, or by tying in a partner with specific skills so that competitors cannot gain access to them. It may constitute one of the strategic games firms play with another. Van Tulder and Junne (1988), for example, refer to the way some companies have a strategy of 'pre-emptive strikes' in signing agreements with attractive partners in order to prevent them partnering competitors. Another example is described by Contractor and Lorange (1988) who provide the example of the way Caterpillar Tractor linked up with Mitsubishi in Japan in order to put some competitive pressure in its home market on their joint competitor Komatsu. The current development of multimedia technology is marked by the grouping of numbers of large firms in the USA, Japan and Europe to develop common standards and thereby exclude other competitors.

Closer customer–supplier links

In some industries, firms have pursued strategies of buying-in rather than making themselves. Major automobile assemblers, for example, increasingly buy-in the technology in their products from component suppliers. Collaboration is a mechanism whereby closer customer–supplier links are established and R&D agendas can be agreed between different partners. Saxenian (1991) describes the way in which, with rapidly changing technologies and shortening product life cycles, traditional customer–supplier links in computer production – typified by the way suppliers are treated as subordinate producers of standard inputs – have been superseded by linkages where suppliers with the status of equals make important technological contributions. This will be examined in more detail in subsequent examination of the Japanese situation. Bessant (1988) suggests that collaboration between customers and suppliers is necessary to deal with the complexity and the high levels of systems integration required for the successful exploitation of Computer Integrated Manufacturing.

   Using his concept of the value chain, Porter argues that the ability to manage the linkages between all the interdependent activities of firms is a decisive source of competitiveness. This includes buyer–supplier links, and also external linkages in R&D. He contends that the exploitation of such linkages within an industry is a critical factor explaining the international competitiveness of that industry (Porter 1990).

Large–small firm links

Small firms are argued to possess advantages over large firms in their ability to respond quickly and flexibly to rapid changes in some technologies and markets (discussed further in Chapter 11). Large firms with their greater resources and marketing and distribution competences, possess advantages over small ones. Combining these advantages of size may be an intent of collaboration.

Declining competitiveness

Mytelka (1991) argues that it is the growing knowledge-intensity of production combined with slow growth associated with economic recession which has stimulated increased collaboration. Economic downturns, she argues, have induced firms to reconsider their strategies for dealing with turbulent markets, and cooperation with other firms – promoting international oligopoly – has been a feature of these strategies.

Positioning, watching and waiting

Collaboration also provides a method by which firms can observe novel technological developments without having to undertake the expense and risk of investing in speculative research. A case in point would be biotechnology, in which many large pharmaceutical companies initially adopted a 'wait and see' strategy to see how the technology developed by collaborating extensively with those entrepreneurial small firms which were investing heavily (Dodgson 1991d). Heaton (1988) quotes a leading Japanese industrialist whose experience of cooperative research led him to argue that its greatest benefit lay in negative research results. That is, it saved the company the costly and time-consuming business of internally pursuing dead-end technology and business avenues.

A key factor affecting a firm's competitiveness is argued to be its positioning in a new technology within a network of suppliers, manufacturers and distributors (Miles and Snow 1986; Osborn and Baughn 1990).

> Early in the development of new products, several feasible designs... from different nations often compete. Such is currently the case with high resolution TV and such was the case with nuclear power and video cassette players. In such areas, knowledge develops rapidly as various firms move to commercialization, consider entering an area, or merely seek to monitor the development of a technology. In short, firms may be seeking to position themselves. They may still be deciding what portions of the technology to keep, whom they will use as suppliers, and how they might successfully market new products... Thus, firms may seek to establish or tap institutional and interorganizational infrastructures and become viable members of a winning network of organizations.
>
> (Osborn and Baughn 1990: 507)

Pre-merger exploration

Before companies merge it may be a prudent step to get to know as much about each other as possible. Collaboration assists companies assess compatibility in organisational structures, management styles, corporate culture and technologies and markets, before full merger goes ahead. For example, the Japanese company Fujitsu worked for many years with the British computer company ICL before taking a controlling interest in it.

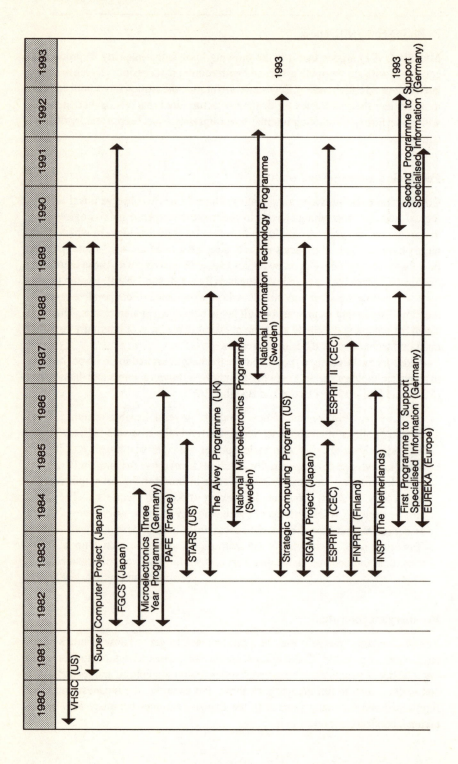

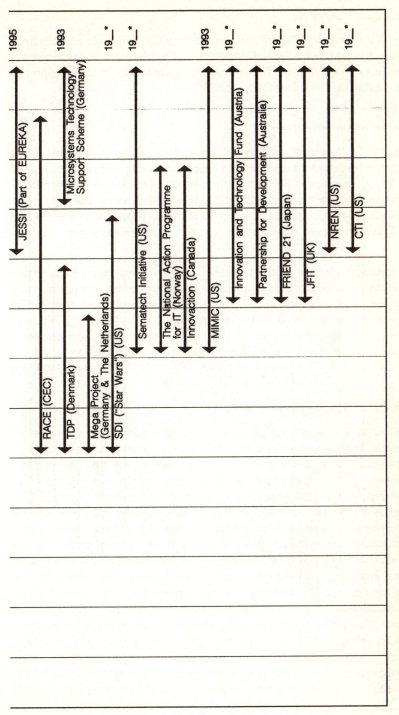

Table 3.1 Major public IT programmes – 1980 to present
Source: OECD

Failure of past mergers

Mergers have a very high failure rate, particularly where they occur between firms in unrelated areas (Porter 1987). Companies which have negative experiences of mergers may choose the alternative option of collaboration. Collaboration is a cheaper option than merger (Mowery 1988).

Problems with alternatives

Mowery (1988) suggests that collaboration provides a valuable alternative to more traditionally used methods of technology buying and selling through licensing and direct foreign investment. He refers to the contractual limitations and high trans-action costs of these alternatives, to the difficulties in 'unbundling' and transferring 'noncodified' firm-specific assets, and to the uncertainty, opportunism and lack of control that complicate international technology licensing agreements. Furthermore:

> The products of a collaborative venture between a US and foreign firm encounter fewer political risks of innovation and foreign marketing. The products of a collaborative venture between a US and foreign firm encounter fewer political impediments to market access in the domestic market of the foreign firm than would direct exports from the US firm. Collaboration also can reduce other political uncertainties that make direct investment in such a market unattractive.
>
> (Mowery 1988: 10)

A PUBLIC POLICY PERSPECTIVE

Few industrialized countries in the world do not have public policies to promote industrial technological collaboration. Table 3.1 lists some of the major collaborative programmes in information technology. One of the first collaborative programmes, and one which in many ways provided a stimulus to others worldwide was the highly successful Japanese very-large-scale-integration (VLSI) project initiated in the late 1970s. In this programme, five computer and semiconductor firms received $360 million public funding to perform very specific technological goals together in a single facility. Collaboration is an important policy target in Japan: roughly 80 per cent of all Japanese government research loans are extended to joint projects (Suzuki 1986, quoted in Levy and Samuels 1991). The European Commission has also played an important role in promoting collaboration, and this will be examined in later sections. There are a variety of reasons why collaboration is promoted by public policies including the belief that collaboration improves innovation potential and the competitiveness of firms and is hence a legitimate aim of industry and technology policy. Pan-European policies to promote collaboration are argued to be a response to the perceived threat of Japanese and USA competition and a belief that European companies are too diversified in their technological activities, too focussed on protected national markets, and commonly replicate research efforts in different countries (van Tulder and Junne 1988).

The DTI, in promoting EC technological collaboration, argues:

> European collaborative research can help both by encouraging industry to carry out more research leading to innovative products and by developing through standards more open markets which increase competition in Europe and hence industry's own competitiveness.
>
> (DTI 1991)

Other, perhaps less laudable, reasons can also be suggested.

With so much pressure internationally to alleviate the burdens of public expenditure, government research laboratories and universities have been increasingly encouraged to cooperate with industry. Collaboration with industrial partners is argued to provide additional funds and market focus for public sector research, and at the same time improve the technological capabilities of firms. Whether it does this, or whether industry actually wants research similar to that it undertakes itself, rather than more basic research, is open to question.

Promotion of 'pre-competitive' R&D rests more easily with those governments whose free-market approaches to industrial and technology policy deter more direct intervention in the behaviour of individual companies. This is particularly apparent in the UK where, as is shown in Figure 2.4, during the 1980s single-company innovation support was almost entirely removed and replaced by support for

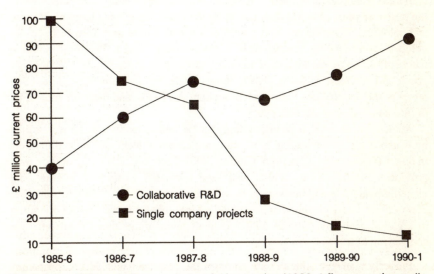

Figure 3.2   DTI expenditure on industrial innovation (1990–1 figures estimated)
Source: House of Lords, 1991

collaboration. In some cases this forced firms, particularly smaller firms into inappropriate collaborations (ACOST, 1990). In other circumstances, large collaborative schemes have been argued to improve the capabilities of 'national' firms to the detriment of foreign competitors. It is, therefore, a highly convenient political policy tool.

Firms are usually attracted by public policies into collaboration in order to obtain the technological and competitive benefits which may accrue, and by the availability of public funds (usually ranging from 30 to 50 per cent of project costs). It should also be noted that these funds are provided for many of the most expensive and risky investments made by firms, and are therefore the more likely to be vetoed in times of economic downturn (see Chapter 7).

In addition to specific collaboration-promoting policies, a range of government trade and industry policies have affected the formation of corporate partnerships. Mowery (1988) refers to a range of policies used in the USA, and against USA firms, to restrict direct foreign access. An example is the use of public procurement and the demand by governments for 'offsets' or the development and production of components for a product by domestic firms in the purchasing nation.

> Government demands for offsets create strong incentives for US producers to involve foreign firms as risk-sharing subcontractors or equal partners in product development and manufacture.
>
> (Mowery 1988: 14)

AN INTERNATIONALIZATION PERSPECTIVE

National technological prowess is nowadays much more evenly distributed than in the post-war years of world domination by the USA. Japan and Germany are also technological superpowers, and countries such as Korea and Taiwan have developed world-leading competences at extraordinary speed. The multi-national nature of technological advantage automatically implies greater trade in technology goods and services. As Soete (1991) shows, trade, foreign direct investment and mergers and acquisitions increased throughout the 1980s, and these are indicators of international inter-dependence.

Markets are believed to be increasingly global, and collaboration is a mechanism by which firms in one trading block can gain access to technologies and markets in others (Porter and Fuller 1986; Ohmae 1990). Technological collaboration may provide a mechanism by which tariff and non-tariff barriers to international technological trade and investment may be circumvented. According to a number of observers we are presently in a situation where global corporations are operating in a global economy (Ohmae 1990). Globalization brings many challenges for companies, but:

> Properly managed alliances are amongst the best mechanisms that companies

have found to bring strategy to bear on these challenges ... In a complex, uncertain world filled with dangerous opponents, it is best not to go it alone.

(Ohmae 1990: 114, 136)

There is a strong body of opinion that collaboration is a feature of the growing globalization of technology (Soete 1991). This is most strongly elucidated in the 'techno-globalism' thesis. This thesis is developed along the following lines. Large companies control the world's technology. The technology strategies of these companies are necessarily international in focus. Large firms internationally access and develop technologies, through collaboration with other companies and some-times in tandem with international scientific efforts, and then manufacture and market them in a multinational framework. The internationalization of private-sec-tor technology and its significance on a world scale has led to 'techno-globalism'. This is argued to have profound consequences for future world economic and technological development.

This is the view of the OECD's Technology Economy Programme (TEP), developed to examine the extent and consequence of international inter-depend-ence. As we shall see in Chapter 9, there are grounds for dispute with the 'strong' analysis of globalization put forward by the TEP study. However, despite very marked differences in national systems of science and technology, and their performance, the broad tendencies of increased international trade, and increasing-ly 'international' strategies of multinational companies which include important elements of collaboration, do provide some partial indications of growing global-ization of technological activities.

At the same time as this is happening, science and technology policies in many industrialized nations commonly share the same concerns and instruments. For example, there is a convergence internationally in policy measures to promote collaboration, generate and regenerate industrial and technological activities in disadvantaged regions, encourage the formation and growth of small firms, and develop infrastructural mechanisms to improve technology transfer (such as science parks) (Rothwell and Dodgson 1992).

Thoughout industry in Europe and the USA, there are enormous efforts to emulate certain of those characteristics of Japanese industry which are believed to underpin its success. Concepts such as Just-In-Time, Total Quality Control and 'lean production' are rapidly being deployed in Western firms. Collaboration is a feature of Japanese industry – horizontally in research programmes, and vertically in subcontracting relationships – and the countries concerned with the rapidly changing nature of international advantage in technology may be attempting to copy the high levels of collaboration in Japanese industry.

CONCLUSIONS

The wide range of forms of technological collaboration described in Chapter 2 is a reflection of the multiplicity of reasons for undertaking it. This chapter has described the many reasons why firms and governments promote technological collaboration. The complexity and uncertainty of the innovation process, its high cost and risks, and the increasing role that technology plays in facilitating inter-firm linkages have all been described as important reasons for collaboration. The competitive and corporate strategy reasons suggested, such as competitor exclusion, positioning and internationalization also centre on the role that technology plays. In subsequent chapters the innovation, corporate, public policy and internationalization perspectives described here are developed in greater depth.

Chapter 4

Theoretical approaches to collaboration

Collaboration excites the attention of analysts and theoreticians from a wide range of disciplines. Essentially this is because it is something which does not fit into well-established knowledge and theories of the firm, or of markets. It is seen as something which lies between the hierarchy and organization of firms and the markets and market forms in which firms operate. As Marceau (1992) puts it:

> The growth of... interlinkages increasingly suggests the 'permeability' of firms' boundaries and that in many cases the effective economic actors are new entities, hard to describe and delineate using the traditional categories of either economists or organisational analysis.
>
> (Marceau 1992)

There exists a wide range of theoretical explanations of collaboration, and the complication this engenders is magnified by the broad lexicon of analytical terms used to describe it. Sako (1992), for example, refers to how it is described as: 'quasi-integration', 'quasi-vertical integration', 'quasi-firms', 'quasi-disintegration', 'visible handshaking', 'invisible handshake', 'invisible link', 'relational contracting', 'intermediate organization', 'network forms of organization', 'dynamic networks', 'industrial networks', 'strategic networks', and 'flexible specialization'; to which can be added: 'innovation networks', 'regional and industrial clusters', and 'disorganized capitalism'!

A review of all these various approaches will not be attempted here (to do so would be tedious and repetitious). Instead, a number of approaches are described which are believed to help explain what collaboration is and why it occurs. These approaches are somewhat crudely separated into those which examine:

1 **changing systems of production**, and the impact of
2 **technological change**; and those that focus on firms and their
3 **economic and competitive relations**, and
4 **organizational learning**.

A number of different analyses are included under each of the four headings.

Demarcation between the various approaches is occasionally arbitrary, as the analyses often overlap. Although they address collaboration, some of the analyses were not specifically designed to focus primarily upon it. The efficacy of the various approaches will be examined in the light of what the previous chapters have told us of the aims and forms of technological collaboration.

CHANGING SYSTEMS OF PRODUCTION

One approach used to explain collaboration is that which sees it as a response to the reorganization and restructuring of industry. For example, Lawton-Smith *et al.* (1991) refer to the *disorganized capitalism* explanation of collaboration based on the Lash and Urry (1987) thesis of the collapse of 'organised capitalism'. In this view, Western companies, unlike those operating in Japan, face increased uncertainty due to the historical loss of control over markets and extensive industrial restructuring. The adhesion to competitive 'free market' principles in industrial policy compounds these uncertainties. Lawton-Smith *et al.* argue that collaboration is a response to this uncertainty.

Marceau (1992) provides a useful distinction in her analysis of the forces 're-working the world', of **chains**, **clusters**, and **complexes** in production. These operate additionally to the influence of the activities of major multinationals. 'Chains' of production refer to the relationships established between 'core' firms and their suppliers and distributors. The chains draw together all those activities from raw material extraction to final product marketing and servicing. The chains are of varying lengths according to the complexity of products, and 'may be conceptualised as the "spines" or "ribs" running through an economy and performing an integrating function such that firms in each are affected by the decisions of others "up" and "down" stream' (Marceau 1992).

Collaboration is an integrating mechanism for these chains, as in the case of closer subcontracting relationships discussed in Chapter 10. In the automotive industry, these linkages are increasingly integrated through the means of 'lean production' (Womack *et al.* 1991) and 'lean supply' (Lamming 1992).

'Clusters' refer to Porter's (1990) analysis of groups of firms, often geographically proximate, which through collaboration and competition internally, continually stimulate product and process innovation and assist external competitiveness. Porter uses a number of examples including the agglomeration of the health-related industry in Denmark; pulp, paper and related machinery in Sweden, and clothing and shoes in Italy. In these examples, the industry has a leading international presence. He describes the mutually supportive links within the cluster, and the way the benefits from these flow vertically and horizontally. He also refers to the ways in which a concentration of rivals, customers and suppliers promotes efficiencies, specialization and innovation.

The way in which industrial districts can form cohesive groupings of innovative and self-supporting firms has been the source of much attention, and is an important element of the *flexible specialization* thesis of Piore and Sabel (1984). Although empirically contestable, this thesis has attracted considerable attention from those who argue that the model is an alternative to the 'Fordist' form of industrial production, and that groups of innovative small firms can in particular circumstances provide a competitive alternative to large-scale organization.

'Complexes' integrate not only firms, but also public-sector bodies, and industry-funded research organizations. The most obvious example of this form of integration lies in the 'military–industrial' complex with its traditionally cohesive inter-relationships. Marceau refers to the example of the construction complex, consisting of groups of large building firms, innovative technologically, working in conjunction with planning authorities, engineering consultants, building research bodies and computing specialists. Gann (1991) provides an analysis of how technological change has extended the construction industry complex. Building contractors are not only using advanced technologies such as CAD, and new management techniques in the organization of construction, but are undertaking R&D into electronic systems, construction robotics, prefabrication and new materials. The demand for 'intelligent', high technology buildings has attracted the entry of diversifying engineering, electronics and telecommunications firms into the construction sector.

These analyses emphasize the systemic nature of industrial change, with extensive inter-connectedness of industrial actors. Marceau (1992), for example, argues that in order to decipher changes in the organization of production it is necessary to examine the relationships of public and private sector, science and technology, large and small firms, and the 'creative tension or stress... which shapes the outcome at any given time and contains the potential for radical change. If the position of any of the key players changes then changes to the whole system are likely' (Marceau 1992).

The chains, clusters and complexes are all inter-dependent. This approach has considerable resonance with many of the factors discussed earlier and underpins many of the ostensible *motives* for technological collaboration.

TECHNOLOGY AND INNOVATION

The *technological primacy* approach includes theories based on the scale, scope and cost of contemporary technologies, whereby firms, in order to deal with the uncertainties of the development and market diffusion of pervasive technologies (particularly information and communications technology), which they cannot individually control, join forces with others, and are increasingly externalizing their technology sourcing and exploitation activities (Granstrand and Sjolander, 1990). As argued in the previous chapter, collaboration plays an important role in technological diversification. This diversification is undertaken not only in products, but also in *capabilities* or competencies (Granstrand *et al.* 1990). Thus a number of innovation and management analyses argue that collaboration is not only of relevance to immediate or envisaged products, but also as a means of developing the capability and flexibility to produce at present unforeseen products (Arnold, Guy and Dodgson 1992).

According to many of the approaches within this technological perspective, a key feature stimulating collaboration is uncertainty about technological development and diffusion. Since Schumpeter, many analyses of technological change have emphasized the *discontinuous* nature of innovation, and the problems this poses for firms (Tushman and Anderson, 1987). This uncertainty extends beyond consideration of technological feasibility, i.e. whether a new product can emerge, to how it will subsequently evolve. Market preferences in new technologies are rarely predictable, and the post-innovation improvements necessary for market success provide additional uncertainty (Rosenberg, 1982; Georghiou *et al.* 1986).

Freeman argues:

> Characteristic of periods of change in techno-economic paradigm is the rise of new firms associated with competence in the new technologies and the strategic re-positioning of many established firms as they try to cope with the rapid structural and technical change affecting their markets and their very existence. If we take into account also the international aspects of production, marketing and technology development, then clearly a period of great turmoil could have been expected in the 1980s, with many new strategic alliances and networks.
>
> (Freeman 1991: 509)

A significant component of this approach is that of *innovation networks* particularly associated with DeBresson and Amesse (1991) and Freeman (1991). Innovation networks are assumed to produce positive sum games for participants in the level of innovation and profits. Their flexibility is useful in dealing with technological uncertainty, and they are helpful in reducing opportunism and in setting technical standards. They are described as the organizational form for the new 'techno-economic paradigm'. This approach helpfully places technological issues and uncertainty centrally in analysis of collaboration, but remains vague about the competitive reasons for, and outcomes of, collaboration. As DeBresson and Amesse (1991) argue, 'As with any powerful concept, that of network is necessarily all-encompassing and therefore vague: it is given different meanings and subjected to various usages' (DeBresson and Amesse 1991: 364).

One of these interpretations of network organization is provided by Imai, who argues it to be:

> a basic institutional arrangement for coping with systemic innovation in the recent technology regime. Networks can be viewed theoretically as interpenetrated forms of market and organization. Empirically they are loosely coupled organization having a core of both weak and strong ties among constituent members.
>
> Cooperative relationships among firms are a key linkage mechanism in network organization.
>
> (Imai 1990: 185)

He argues that Japanese business organization is evolving from its traditional, formal *Zaibatsu* and *Keiretsu* business groups to a more adaptable and flexible network organization (see Chapter 10).

The concept of innovation network is perhaps given its greatest credence by Saxenian (1991) in her work on computer systems in Silicon Valley. She argues:

A network of long-term, trust-based alliances with innovative suppliers represents a source of advantage for a systems producer which is very difficult for a competitor to replicate. Such a network provides both flexibility and a framework for joint learning and technological exchange.

(Saxenian 1991: 430)

Apart from facilitating information exchange, networks in Saxenian's view also promote innovation by encouraging specialization and joint problem solving, spreading costs and risks, and foster the application of new technologies because they encourage new firm entry and product experimentation.

However, this approach does have its critics. Hobday (1991) contends that in the light of continuing large Japanese firm predominance, Saxenian's analysis, rather than describing a model for industrial success, is one which has distinct limitations, particularly as the firms in the networks she describes have little capacity for manufacturing and thus maximizing the returns from their innovation. Hobday suggests that the innovative network may only have advantages at the early stages of a new technology's development.

Saxenian also observes that traditional arms-length relations persist, for example, with suppliers of such commodity inputs as raw materials, process materials, sheet metal, and cables. Furthermore, many of these inputs come from outside the region, and often from Japan.

The innovation network approach has some semblance with Porter's (1990) analysis of industrial clusters. Porter, however, places greater emphasis on the role of large firms in the network and its competitive rather than cooperative nature.

A number of analyses of collaboration refer to the way the intent, focus and nature of partnership vary with product and technology *life cycles*. Cairnarca, Colombo and Mariotti's (1992) approach was described in Chapter 2. Life cycle analysis of collaboration is also a major feature of Kogut (1988), Mody (1989 and 1990), James (1989). Uncertainties at the early stage of product or technology life cycles are argued to lead to collaboration between firms in R&D. As products and technologies mature, competition is driven by cost considerations, and collaboration is used to link assemblers and suppliers in novel ways to reduce cost. When products and production technologies are mature firms seek collaboration in order to rejuvenate them or to find alternatives. Life cycle analysis is, of course, much more sophisticated than this simple sequential model. Firms begin collaboration in production in the early stages of life cycles, and begin collaboration in R&D long before products are mature. Furthermore, firms are argued to have the ability to influence the speed of life cycle changes by influencing the behaviour of competitors. An important aspect of the life cycle approach is the way in which it naturally emphasizes the termination of collaborations, a factor omitted from many other approaches.

The nature of *intellectual property* – private, public or 'leaky' i.e. that which can only be privately appropriated for a short period – and the institutions which encourage the creation of each provides the focus of Ouchi and Bolton's (1988) analysis of collaboration. Leaky property has the advantage of providing incentive for its creation (short-term monopoly power), and public good (knowledge ends up in the public domain). Collaboration, the authors argue, provides the institutional form needed to produce leaky property which is necessary in competitive world markets. This is not only an aim of public policy. Nelson (1988), when discussing the way in which certain industries 'pool' patents, argues:

> Such arrangements reflect an apparent agreement among a group of firms that they are all better off if they make a common, big pool of at least some of their technological knowledge, than if they all try to keep their individual pools strictly private. It seems that, within limits at least, rivalous companies can and do recognize the 'public good' properties of technology.
>
> (Nelson 1988: 318)

The approaches described here are broadly complementary to those examining changing systems of production, although placing greater emphasis on overtly technological issues, particularly the question of uncertainty, and point to one of the key elements of Marceau's creative tension and stresses within industrial systems. They move beyond analyses of motive and consider *processes* of collaboration inasmuch as they begin to analyse the community nature of firms in collaboration and the way in which they alter over time. They suffer, particularly in comparison with the following approaches, in the clarity and depth of analysis of the motives of technological collaboration. Technological collaboration is a means to an end – long-term competitive survival, growth and profitability – not an end in itself as much of this analysis implies.

ECONOMIC AND COMPETITIVE RELATIONS

The *transaction costs economics* approach in analysing collaboration is most closely associated with Williamson (1975, 1985). For Williamson, transactions are the basic element of analysis of all economic institutions and, to a great extent, efforts to economize on transaction costs determine organizational behaviour. Transaction costs include the costs of searching, bargaining, deciding, monitoring and enforcing transactions. 'Transaction costs are economized by assigning transactions (which differ in their attributes) to governance structures (the adaptive capacities and associated costs of which differ) in a discriminating way' (Williamson, 1985: 18).

The governance structures selected may range from classical market contracts to highly centralized, hierarchical organization, with mixed forms of market and firm organisation in between. Williamson contends that hierarchies generally are more economical in transaction costs, but because of increased problems with size (greater bureaucratization) contracts may, on the basis of some continuing commitment between partners, economize on transaction costs.

Transaction costs are affected by a number of other issues, including: asset specificity, bounded rationality/opportunism, and uncertainty. Asset specificity refers to those assets which are particular to individual transactions. They may relate to site specificity, physical asset specificity, human asset specificity and dedicated assets (specific to particular buyers). Without asset specificity, according to Williamson, contracts would be vastly simplified. Individual actors in transactions are assumed, as in Simon's (1961) sense, to intend to behave rationally but in fact their rationality is limited. Managers are assumed to be guileful and behave opportunistically. Uncertainty occurs where information is lacking or is unequally distributed, and is common in many transactions, and particularly related to technology. Alliances are seen as a form of hierarchy, which when transactions are costly, complex and difficult to specify, are a more efficient form of governance structure than contracts in markets (Williamson 1985).

Transaction costs economics has a very broad currency. Langlois (1989) argues that almost all modern economic theories of vertical integration are transaction costs explanations. Some, like Langlois' own, modify or extend the analysis, and include such factors as the size of markets, their rate of change, and incorporate past history. Most, however, share assumptions about base human behaviour which is probably the weakest element of the thesis. Collaboration is a means by which 'leakage' of proprietorial information can be restricted to small numbers of firms, rather than having to display this information for broad consumption (and potential replication) in markets. The assumption of opportunistic managerial behaviour is carried into the relationship between firms in collaboration. Collaborative efficiency is assumed to be achieved when each partner can limit others' behaviour by means of a 'double-hostage system'.

This approach is antithetical to those many approaches which emphasize the importance of high levels of trust between partners (discussed in Chapter 12). The instrumentality of the approach excludes questions of organizational power and the inter-personal relationships which, as will be discussed in later sections, provide some of the reasons for collaboration, and are so central to its process and outcome. The great strength of this approach is the simplicity of its analysis and predictive capability: the option of market, hierarchy or intermediate governance structure is decided primarily according to transaction cost minimization. This is singularly inappropriate, however, in explaining the learning and capability-building motives of technological collaboration. Such issues cannot be incorporated within considerations of asset specificity and transaction cost economizing, and therefore have limited value in this case.

Much of the *strategic competitive analysis* approach is concerned with structural questions of how the firm relates to its competitors. The approach focusses on the way that industrial structures (affected by their level of concentration and competition, existence of scale economies and other entry barriers, and general levels of technological change) influence the behaviour of firms (Porter 1990). In this approach collaboration is seen as a means of shaping competition by improving a firm's comparative competitive position through raising entry costs, increasing price performance differentials, or by reducing uncertainties by encouraging mutual dependencies. Globalization is an important element of Porter and Fuller's (1986) analysis, inasmuch as it stimulates structural change. For Hamel, Doz and Prahalad (1989), collaboration needs to be viewed in a competitive power perspective. Collaboration is a continuation of competition, and should be seen, as Harrigan (1988) sees it, as a transitional stage in firm positioning. This approach is helpful in introducing the notions of competitiveness and market power into the discussions.

One element of this approach is *game theory*. This points to the way that firms send signals to one another, informative and misleading, which are designed to influence other firms' behaviour. The formation of collaborations may be one such signal. Axelrod's (1984) work with the 'Prisoner's Dilemma' game may have some implications for strategic management of collaboration inasmuch as it reveals the benefits of reciprocity and of cooperation when expectations of partners are long-term.

A distinctive, but related, approach in this tradition is provided by Miles and Snow (1986) in their conception of *dynamic networks*. This approach contends that future competitiveness will depend on the ways in which firms interact with one another. Unlike the approach to 'innovation' networks discussed earlier, its focus is not primarily technological. In arguing that future industrial structures will be disaggregated, and market transactions will replace previously internalized activities, this view suggests the growing importance of collaboration. Indeed, it argues that the ability to operate in the network will be a key source of competitiveness. The reasons why networks are formed, and considerations of their comparative performance, are rather vague in this approach. Rather more attention to these questions is provided by Jarillo (1988) who refers to *strategic networks*, which are 'long-term, purposeful arrangements among distinct but related for-profit organizations that allow those firms in them to gain or sustain competitive advantage vis-à-vis their competitors outside the network' Jarillo (1988: 32).

He argues that a network's effectiveness can lie in technological reasons (achievement of lower external than internal costs), and the possibility of lowering transaction costs. Its efficiency can be measured in long-term profits compared to going alone. Focussing on the former rather than the latter, he argues that networks provide an opportunity for joint value creation; because of specialization they reduce final total costs, and as networks are typified by high levels of trust, transaction costs are also reduced. Furthermore, Jarillo suggests the mechanism for network creation lies in the catalytic role of a 'hub' firm cognisant of the advantages of network organization.

The resource-based perspective, associated with Teece and Pisano (1987) and Mowery (1988) integrates a number of the insights of other approaches. It sees firms as 'bundles of resources', and refers to the comparative cost efficiencies in transaction structures. It also emphasizes the imperfections of transfer mechanisms of knowledge, technology and other assets. For Teece (1986), full commercial rewards from innovation can only be achieved if firms can access 'complementary assets' such as competitive manufacturing and distribution and marketing. These assets can be linked to R&D by means of arm's-length transactions; vertical integration; and collaboration. Arm's length transactions can have high costs, but are useful when technology is codified, discrete (non-systemic) and relatively simple. Vertical integration limits transaction costs, but prevents the access of specialisms in other firms. Collaboration allows these specialist skills to be accessed, and can allow complex and tacit knowledge to be transferred, and technology to be 'unbundled'. Important in this analysis is the question of appropriability: how firms protect and utilize their intellectual property.

These latter approaches further help our understanding of the motives and processes of technological collaboration. They usefully highlight the need to consider the competitive and profitability aims of collaboration, the reasons for and constraints on resource, exchange, and the continuing dilemma between firms' competitive aims and cooperative means.

COLLABORATION AND LEARNING

Many of the approaches listed above tend to offer an 'instrumental' explanation for collaboration, inasmuch as firms are assumed to respond to cost, competitive power, resource availability and technological imperatives. They commonly disregard 'behavioural' explanations, in the sense that firms and individuals can make highly variable strategic choices in response to similar stimuli. Another approach, which emphasizes the importance of organizational and technological 'learning', can more fully include individual and individual firm behaviour in its analysis. Some of the approaches to learning and collaboration are described, and reasons for the need to learn are examined, as suggested by literature within an organization studies and a strategic management perspective.

There exists a growing literature on the relationship between learning and collaboration (see, for example, the chapters in Contractor and Lorange (1988) by Westney, Kogut, Lyles and Hakansson and Johanson). For Kogut (1988) joint ventures are 'vehicles by which knowledge is transferred and by which firms learn from one another'.

The primary ascribed *motive* for learning through collaboration is to deal with technological and market uncertainty (Ciborra 1989, 1991; Mody 1990). Mody describes learning as a particularly strong motive for forming and sustaining alliances: 'The purpose of the alliance is to discover whether complementary capabilities make sense from a technological and/or market perspective' (Mody 1990: 11).

Ciborra is perhaps the most enthusiastic proponent of collaboration as learning process. He argues that alliances are the institutional arrangement that most efficiently allow firms to implement strategies for organizational learning and innovation. 'Alliances... provide, through new insights and successful restructuring of organizational problems, a shortcut to radical change by-passing organizational inertia and deadlocks' (Ciborra 1989: 12).

Ciborra's analysis is based on the premise that collaborations reduce uncertainty by improving predictability of technological development, and that they are a means of reducing the *transition costs* of firms transferring strategies. Both these assumptions are questionable, but the latter one in particular is problematic as it appears to ignore the continuing importance of firms' 'knowledge bases' or 'firm-specific competences'. Networked R&D is seen to be the primary mechanism for dealing with turbulent technological change. Firms' internal R&D efforts, and the ways in which firms can control the pace of technological development appear to be relegated in importance contrary to the findings of the majority of studies into technological development (Freeman 1982).

Learning is argued to be an important outcome of collaboration in this literature, and these outcomes can be manifold. To quote Ciborra again:

> The alliance brings into the corporation new expertise concerning products, marketing strategies, organizational know-how, and new tacit and explicit knowledge. New management systems, operating procedures and modifications of products are the typical outcomes of this incremental learning. A firm can learn how to set up and fine-tune alliances per se. The result of such learning is the institutionalization of the organization's rules and routines aimed at managing alliances.
>
> (Ciborra 1991: 59)

A valuable contribution made within the learning approach to collaboration is the emphasis placed by some of its proponents on the necessity for adaptability, change and conscious choice in collaboration. As Contractor and Lorange (1988) argue, the strategic rationales prevailing when a cooperative venture was formed may shift over time. They argue that the erosion of the fundamental strategic rationales may come from external or environmental sources (such as technological obsolescence) or internal sources 'such as when one partner learns from the other, and the other partner has nothing to contribute'.

Learning and adaptability are necessary in collaboration because:

(a) the bargaining power of partners varies over time (Kogut 1988; Doz, 1988);
(b) the original reasons for the formation of collaborations may become obsolete over time (Harrigan 1986);
(c) initial agreements surprisingly often focus attention on the wrong sets of issues (Lyles, 1988).

Doz and Schuen (1988) argue that there are three learning processes in continuing partnerships: learning about the partner, learning about the task, and learning about outcomes.

(a) Learning about the partner, they argue, is important as the most relevant organizational information is tacit in nature.

(b) Learning about the task relates to the way objectives and requirements become clearer following the establishment of partnerships.

> the very conditions that motivate the partnership often make the task or project envisioned difficult to plan precisely at the outset. The partners are dealing with areas unfamiliar to them in one or more aspects, the environment may be hostile, the task not yet well specified and neither of the partners have the full complement of capabilities or skills to complete the task on their own. Through the partnership, the partners learn more about the task requirements and clarify the achievable outcomes and their timing.
>
> (Doz and Schuen 1988: 11)

(c) Learning about outcomes.

> Partners initially start off with a too rosy picture of the expected outcomes. This over-expectation can result from the bargaining process, where each partner is encouraged to oversell its advantages and capabilities and undersell its weaknesses.
>
> (Doz and Schuen 1988: 12)

They also point out how much bargaining process takes place within firms, as the advocates of partnership tend to oversell the advantages of collaboration to top management.

Dodgson (1991d) argues that it is the differential ability of firms to learn quickly about technological opportunities that has been responsible for the important, complementary and changing roles of large and small firms in the evolution of biotechnology. Pucik (1988a) argues that in collaborative partnerships: 'Benefits are appropriated asymmetrically due to differences in the organisational learning capacity of the partners. The shifts in relative power in a competitive partnership are related to the speed at which the partners can learn from each other' (Pucik 1988a: 80).

In his study of Japanese and US collaborations Pucik found distinctive advantages in the approach of Japanese firms which had developed a systematic approach to organisational learning:

> Japanese firms put in place managerial systems that encourage extensive horizontal and vertical information flow and support the transfer of know-how from the partnership to the rest of the organization. The policies guiding the management of human resources at all levels and functions constituted a vital part of such a learning infrastructure.
>
> (Pucik 1988b: 81)

Aoki (1988) discusses the advantages Japanese firms have over American in their ability to offer incentives to information exchange and learning:

The A-firm emphasises efficiency attained through fine specialization and sharp job demarcation, whereas the J-firm emphasises the capability of the workers' group to cope with local emergencies autonomously, which is developed through learning by doing and sharing knowledge. In the former, the operating task is *separated* from the task of identifying and finding necessary expedients to overcome and prevent emergencies, whereas in the latter the two tend to be *integrated*.

(Aoki 1988: 16)

Pucik's emphasis on the human resource aspects of learning through collaboration is, as seen earlier, supported by other studies. Importantly, he also points out that traditional planning systems cannot assign a financial value to learning activities, and organizational learning is therefore left unfunded. The costs of learning are immediate, the benefits likely to be long-term.

These approaches to learning do not tell us a great deal about the factors which influence and constrain firms' needs and desires to learn. Two approaches which attempt to do so – from an organization theory and strategic management perspective – are described below.

Within organization theory there has been a range of explanations for the problems firms have in learning to do new things. Argyris and Schon (1978) develop a three-fold typology of learning which they describe as single-loop, double-loop and deutero-learning.

Organisational learning involves the detection and correction of error. When the error detected and corrected permits the organisation to carry on its present policies or achieve its present objectives, then that error-detection-and-correction process is single-loop learning. Double-loop learning occurs when error is detected and corrected in ways that involve the modification of an organisation's underlying norms, policies and objectives.

(Argyris and Schon 1978: 3)

Organizations need to learn how to carry out single and double-loop learning. This Argyris and Schon call deutero-learning.

When an organisation engages in deutero-learning its members learn about previous contexts for learning. They reflect on and inquire into previous episodes of organisational learning, or failure to learn. They discover what they did that facilitated or inhibited learning, they invent new strategies for learning, they produce these strategies, and they evaluate and generalise what they have produced.

(Argyris and Schon 1978: 27)

Argyris and Schon note the ways in which organizations create learning systems which inhibit double-loop learning. Primary inhibitory learning loops are a self-reinforcing cycle in which errors in action provoke individuals to behaviours which reinforce those errors. Secondary inhibitory loops are group and inter-group dynamics which enforce conditions for error through, for example, ambiguity and vagueness. They contend that 'Organizations tend to create learning systems that inhibit double-loop learning that call into question their norms, objectives, and basic policies' (Argyris and Schon 1978: 4).

In their research they found no organization which double-loop learned. They do not, however, discount its ever occurring. Morgan (1986) also analyses learning inhibitors. He describes how departmental structures focus the attention of their members on parochial rather than organization-wide problems; how systems of accountability frequently foster defensiveness in attitudes; and how there is a gap between actors' rationalized statements of what they do and what actually occurs.

A similar argument is developed by March (1991) who distinguishes 'exploration' and 'exploitation' in organizational learning. Exploitation involves the refining and extension of existing technologies and competences. Exploration is experimentation with new alternatives. As the returns of exploration are uncertain, long-term and often negative, March argues that organizations have a 'tendency to substitute exploitation of known alternatives for exploration of unknown ones'.

Another issue of importance discussed by Hedberg (1981) is 'unlearning':

> Understanding involves both learning new knowledge and discarding obsolete and misleading knowledge. This discarding knowledge – unlearning – is as important a part of understanding as is adding new knowledge. In fact, it seems as if slow unlearning is a crucial weakness of many organizations.
>
> (Hedberg 1981: 3)

In some of the management literature on learning in Japanese firms, the ability of these firms to unlearn has been described as a reason for their comparative innovative efficiency (Imai *et al.* 1985).

A particularly valuable approach to strategic management has recently emerged which focusses on learning in individual firms, and the contextual factors which influence it. Teece, Pisano and Schuen (1990) develop an approach to strategic management which they describe as '*dynamic capabilities*'. In some senses it attempts to incorporate 'instrumental' and 'behavioural' analysis in the senses used earlier. The strategic management literature has long been aware of the problems of organizational resistance when strategies are amended (Ansoff 1968). Porter (1990) describes the way that

> change is extraordinarily painful and difficult for any successful organization. Complacency is much more natural. The past strategy becomes ingrained in organizational routines. Information that would modify or challenge it is not sought or filtered out. The past strategy takes on an aura of invincibility and becomes rooted in company culture. Suggesting change is tantamount to disloyalty. Successful companies often seek predictability and stability. They become preoccupied with defending what they have, and any change is tempered by the concern that there is much to lose.
>
> (Porter 1990: 52)

The focus of the dynamic capabilities approach, which builds on a variety of previous theories of strategic management (including Porter's), are 'the mechanisms by which firms accumulate and dissipate new skills and capabilities, and the forces that limit the rate and direction of this process'. Based on traditional theory of the firm, Teece *et al.* argue that competitiveness is derived from the ability to reproduce over time distinctive organizational competences. Central to this approach is the question of learning. Building on recent work on theory of the firm, they also encompass concepts such as path dependencies, technological opportunities and timing, transaction costs, complementary assets and selection environments.

Path dependencies refer essentially to the way that a firm's history helps define and direct future activity. Technological opportunities and timing are a function of firms' R&D budgets and project selection, developments in the science base, and the way large firms can assist the direction of technological development. A salient concept in this regard is that of 'technological trajectories' (Nelson and Winter 1982). In the circumstances of specific firms these refer to the way that technology incrementally develops within a firm, and reflects the technology's relationship with endogenous factors, such as the firm's cumulative learning abilities; and exogenous factors, such as market pressures. Timing also emphasizes the importance of chance, identified by Arthur (1989) as being so important in economic events. Arthur also highlights the problems of path dependency, particularly when firms and products are 'locked-in' to inferior paths of technological development. Under such circumstances, the need for 'unlearning' is very apparent.

Transaction costs, as seen above, introduce the question of relative cost efficiencies of differing organizational forms. Selection environments essentially refer to the degrees of freedom companies have in which to operate, and include factors such as the extent of competition, capital constraints, and how path dependent firms are.

This broad-based theory has some support in the more empirically-based work of Pavitt (1991) on innovation in large firms which again has learning as a focus. Innovative activities, he argues, are firm specific, cumulative, differentiated and highly uncertain. Like Teece, Pavitt argues the central importance of tacit knowledge obtained through experience and how this, amongst other things, ensures that what firms did in the past will condition what they do in the future. The range of technological options open to firms is strongly dependent on their accumulated skills in proximate technologies, so the nature of technological opportunities and threats facing firms varies according to their principal activities. He finds that the technological diversification of firms tends to follow specific paths according to their principal sector. For example, chemical firms are increasingly diversified upstream, downstream and horizontally; while mechanical, instruments and electrical/electronic engineering firms are diversifying horizontally (Pavitt, Robson and Townsend 1989).

Unlike Teece *et al.*, Pavitt refers to the organizational challenges facing large firms, including the need for collaboration between professionally and functionally specialied groups; questions of the viability of multi-divisional form structures; and considerations of the advantages of centralization versus decentralization.

In addition to the internal organizational factors which inhibit higher level learning and encourage introspection and conservatism in learning there are also contextual, firm-specific constraints to learning. These factors can be argued to have profound implications for firms during periods of turbulent technological change. As will be argued in Chapter 13, collaboration can provide a method by which these learning constraints are overcome, and can provide a stimulus to 'higher level' learning.

The attempt made by these approaches to integrate many of the contextual factors which influence firm behaviour applies also to their decisions about collaboration. Technological factors, resource availability, the nature and form of knowledge, relative costs and corporate history and strategy are likely to affect the propensity to collaborate. When these are supplemented with the insights into the organizational problems faced by firms in their learning activities, then these theories provide a reasonably robust, if very broad, view of the reasons for the motives and processes of collaboration. Furthermore, as future chapters will argue in greater depth, learning is also an important *outcome* of collaboration. The learning approach to collaboration is a very broad theoretically but it is necessary because of the complex multiplicity of factors influencing technological collaboration.

CONCLUSIONS

In this brief review many of the theoretical approaches applicable to technological collaboration have been described. In relation to the previous discussion of the reasons why firms collaborate in their technological activities some of these analyses have been argued to be explanative, others less so. It is useful to place the turbulence and uncertainty facing firms' operating environments within the big picture of changing industrial structures, and envisioning the changes within chains, clusters and complexes helps understanding of the systemic nature of this change. Those approaches focussing primarily on technological change usefully emphasize the extent of the uncertainty and turbulence facing firms, and suggest the way in which the cyclical nature of technology development and diffusion may impact the motive and process of collaboration. It furthermore begins to delineate some of the cooperative elements of relationships between firms necessary for technological development.

The approaches focussing on economic and competitive relations valuably direct analyses to the key motive of technological collaboration: long-term corporate survival and growth. They introduce the question of power between firms, and how dominant positions may be maintained. However, the transaction cost approach to explaining technological collaboration was found to be very limited in a number of respects. First, it discounts many of the important characteristics of collaborations discussed in the literature, and further elucidated in later chapters, of inter-personal and organizational trust. Second, it cannot account for those aspects of collaboration which stimulate and facilitate learning in firms, and which enable the development of broad technological capabilities.

The organizational learning approach was argued to have the advantage of analysing motive, process and outcome of technological collaboration. By looking at the problems within firms in dealing with novelty, the necessity of learning because of the many reasons described in the other approaches, is strongly emphasized. Some of the characteristics of this learning process are analysed, including higher and lower levels of learning, and unlearning. The questions of learning through collaboration will be returned to in later sections when attention is directed to the outcomes of collaboration.

Chapter 5

Collaboration and innovation – the case of biotechnology

Innovation is critical to the competitiveness of firms, industries and nations. Some of the ways collaboration assists the innovation process will be addressed in this and the following chapter. Inter-organizational linkages are very important for innovation. This chapter describes the role of these linkages in biotechnology. It argues that firms link together for many of the reasons suggested in earlier chapters, such as cost and risk sharing, and the merging of complementary capabilities. Public policies sometimes play a role in stimulating collaboration, as happened in biotechnology. A case study of a typical collaboration in biotechnology is presented, and some of the characteristics of collaboration are described, such as the extent of the management task involved, and the role that trust plays in creating adaptable and long-term partnerships.

COLLABORATION AND THE INNOVATION PROCESS

Innovation can be highly radical and completely change existing ways of doing things. This is rare, however, and it is more often incremental, improving products or processes. Freeman (1982) describes it as the technical, design, manufacturing, management and commercial activities in the marketing of a new (or improved) product or the first use of a new (or improved) manufacturing process or equipment.

It is a complex and iterative process. Some of its complexity can be seen in Rothwell and Zegveld's (1985) representation of the innovation process in an 'interactive model', shown in Figure 5.1. They regard the process as being:

> A logically sequential, though not necessarily continuous process, that can be divided into a series of functionally distinct but interacting and interdependent stages. The overall pattern of the innovation process can be thought of as a complex net of communications paths, both intra-organizational and extra-organizational, linking together the various in-house functions and linking the firm to the broader scientific and technological community and to the marketplace. In other words the process of innovation represents the confluence of technological capabilities and market needs within the framework of the innovating firm.
>
> (Rothwell and Zegveld 1985: 50)

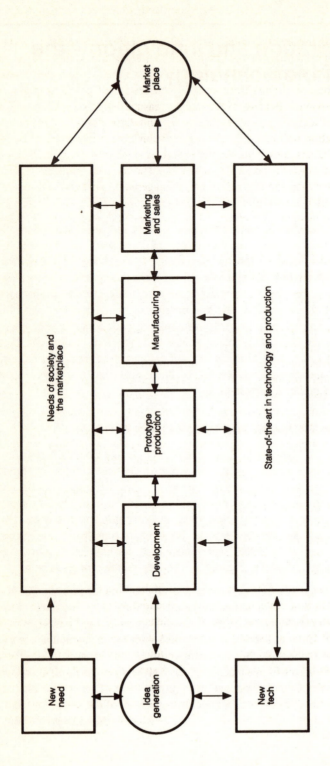

Figure 5.1 The 'coupling' model of innovation
Source: Rothwell and Zegveld 1985

The firm is, therefore, the medium by which market and societal needs and technological possibilities merge into innovation. While innovation occurs within firms it is itself a response to wider pressures outside the firm and does not occur in firms in isolation. Inputs from external sources are of crucial importance to successful innovation. Because of some of the reasons described in Chapter 3, such as the nature of technology – its tacitness and cumulativeness – and the problems of communication between different organizations, these inputs are often formalized into collaborations. This chapter examines some of the ways collaboration between organizations encourages innovation. The focus is on technological innovation, rather than the organizational, management, financial and marketing innovations which are so often also needed to ensure success. It concentrates on the science and technology elements of the process rather than the market ones.

COLLABORATION IN THE EVOLUTION OF A NEW TECHNOLOGY: THE CASE OF BIOTECHNOLOGY

Biotechnology is one of the most important new technologies to have emerged in the last two decades, and its development is very revealing of the importance of collaboration in the process of innovation. Biotechnology is defined as the industrial use of a number of new techniques – recombinant DNA, monoclonal antibodies (MAbs), and bioprocessing – on living organisms (or parts of organisms) to make or modify products, to improve plants and animals, or to develop microorganisms for specific uses (OTA 1991). In some ways biotechnology is a radical technology, inasmuch as it has the potential to be highly disruptive of existing ways of doing things (Hamilton *et al.* 1990). More often nowadays, as its potential is more clear, it is seen as a technology which complements existing ways of doing things without actually replacing them. Its evolution is highly illustrative of the role of collaboration in innovation.

The basic scientific breakthrough which began the evolution of the 'new' biotechnology, the first cloning of a gene, occured in 1973. In less than twenty years this technology has spawned an enormous research effort, affected the structure of a number of industries, and is beginning to produce significant new products in the large pharmaceutical, agriculture and chemical markets (OECD, 1989). According to one estimate, biotechnology will be a $70 billion business worldwide by the year 2000 (Dibner 1991). Collaboration has played a major role in the short history of biotechnology. It has occured extensively between universities and industry (Kreiner and Schultz 1990; Senker and Sharp 1990) and between large and small firms in industry (Arora and Gambadella 1990). High levels of collaboration have occurred between these various actors because of the differing competences within them, and the demand for speedy innovation requiring their combination (Dodgson 1991d).

A number of studies have shown the high level of collaboration between the science base of universities and research hospitals, large firms, and a new category of small firm established on the basis of biotechnology: the dedicated biotechnology firm (DBF) (Pisano *et al.* 1988). The proportion of these collaborations which focus on R&D is, like the other studies shown earlier in Table 2.2, around one-third (Pisano *et al.* 1988; Hagedoorn and Schakenraad 1990). Yarrow (1988) found that on average the ten largest USA pharmaceutical companies had three partnerships with DBFs. The extensive external linkages of one DBF, Centocor, are described in Freeman and Barley (1990).

Table 5.1 shows the number of international collaborations enjoyed by five leading USA DBFs.

DBFs are significant players in biotechnology in the USA. They are estimated to employ 73,000, have revenues of $12 billion and spend $3.6 billion on R&D. However, as the source of these statistics shows, compared to DBFs, large corporations involved in biotechnology have on average 250 times the personnel, 100 times the R&D budgets, and 400 times the revenues (Dibner 1991).

Collaboration was a feature of biotechnology right from the start simply because it emerged in the science base rather than in the laboratories of large companies. The two major scientific breakthroughs which led to the development of biotechnology – in recombinant DNA and MAbs (cell fusion) – occurred in the universities of Stanford in the USA and Cambridge in Britain. The academics' skills in 'genetic engineering' were very novel, leading in the British case to the award of the Nobel Prize. Furthermore, concern over the consequences of tampering with genetic structures initially led to very strict controls over where experiments could take place, and there were few laboratories in the world where such conditions were met (Hall 1987).

The earliest scientists in the area, with their new skills, unsurprisingly became highly attractive to large companies working in areas such as pharmaceuticals where biotechnology potentially could make a profound impact, and also to the growing number of venture capitalists who, having seen the fortunes made in the electronics revolution, were hoping to make their own in firms created to exploit the latest new technology. The enthusiasm for working with these scientists led to their recruitment into industry and to research contracts and collaboration. Recruitment tended to be into small, start-up firms rather than into large companies as this sort of firm enjoyed working conditions and practices more akin to those found in academia, and also the stock options of small, fast-growing companies proved very

Table 5.1 Number of international agreements by biotechnology companies

	(1989)
Biogen	15
Chiron	16
Genentech	13
Genetics Institute	9
Centocor	8

Source: OTA 1991

attractive (Hall, 1987). Collaboration between the university and research hospital departments and industry has continued at a high level in efforts to transfer the basic scientific skills and knowledge into industrial firms (Pisano *et al*. 1988; OTA, 1991). Dibner (1991) in a study of 742 USA DBFs found that 70 per cent have alliances with US universities; 40 per cent with European universities; and 12 per cent with Japanese universities.

Superficially, the development of biotechnology has a number of similarities to Schumpeter's entrepreneurial model of innovation. In this stylized model, science and inventions (which are developed outside of industry) attract the attention of entrepreneurs who then induce new investment in novel technologies, and lead to new production patterns and changed market conditions. The attractiveness of the high profits from the entrepreneurial investment causes expansion or 'swarming' of a large number of new entrant firms. These new firms eventually supersede the old, which have to adapt or go out of business. Eventually, with new entrants, profits decline and numbers stabilize.

Similarities exist with this model in the way that biotechnology emerged within the science base. It accords with the way that, on the basis of new forms of capital, entrepreneurs established new, start-up firms, numbers of which rapidly expanded but then levelled off and contracted (see Dodgson 1991d).

There are, however, major differences. First, large, established firms have played a major role in the development process (without the initial research being undertaken by them as would be predicted by Schumpeter's alternative 'large firm' model of innovation). Dibner (1991) estimates that 142 major corporations have established significant biotechnology efforts in the USA. Second, no large firms have been displaced by new entrants. Third, the number of new entrants has levelled off because of limited incentives: profits have been low or non-existent.

Biotechnology has, in fact, evolved in circumstances where entrepreneurial start-ups have emerged and grown rapidly *and* large firm hegemony has continued. As in the case of the development of semiconductors, innovation has depended on the 'dynamic complementarity' of large and small firms (Rothwell 1983).

One of the major reasons why the dominance of large firms continues in biotechnology is the inability of DBFs to access 'complementary assets' (Teece 1986). In biotechnology these complementary assets, which are necessary for firms to attain full returns from innovations, may include competitive manufacturing, marketing and distribution networks, and the ability to deal with the regulatory procedures involved in getting new products on to the market. Initially, as biotechnology began to emerge from the science base in the early 1980s, DBFs possessed comparative advantages in R&D which had been transferred from the universities. They had no manufacturing expertise, marketing networks, nor experience with regulatory issues. Large firms had limited R&D skills and difficulties in putting together the appropriate mix of skills in biotechnology, but they had the complementary assets DBFs needed. These distinctive advantages provided the basis for collaboration.

Over time, the picture has become less clear. Large firms have developed their own skills in biotechnology R&D. DBFs have moved into manufacturing and developed knowledge and expertise of regulatory procedures, and some have developed marketing capabilities. Yet extensive collaboration continues between the two groups of firms. One reason for this is the sheer level of resources required to enable many biotechnology products to proceed through the regulatory process. In the pharmaceutical area – in which the majority of the world's DBFs operate – it can cost around $200 million to develop a new drug. This constraint is so high that it led the most successful DBF in the world, Genentech, to seek to be acquired by a large firm, and to the merger of two leading DBFs: Chiron and Cetus. Many DBFs have depended on large firm finance from R&D contracts and collaborations; technology has been exchanged in some collaborations; and large firms frequently provide the 'market' of potential licensers of DBFs' products. Another reason for the high level of collaboration lies with the differential organizational abilities of DBFs and large, established firms (Dodgson, 1991e).

As it emerged, biotechnology did not accord with large pharmaceutical firms' existing structures or drug discovery procedures. Most large pharmaceutical firms' R&D efforts were organized into 'Chemistry' and 'Biology'. The new skills of molecular biology did not fit well into these structures. Traditional drug discovery procedures involved the search for new substances and then screening them for potential uses. Biotechnology, alternatively, has the potential to use the skills of genetic engineering to 'design' drugs with particular properties.

While biotechnology has proved a turbulent technology for large firms it has not led to their demise and replacement by new firms, as in Schumpeter's sense, for a number of reasons. First, the technology is still in its infancy. Second, new generations of technology rarely succeed older vintages overnight – there is a period of co-existence (Pavitt 1987). Third, large firms have begun to build their own skills, often in tandem with DBFs in collaborations, but also increasingly through acquisition of DBFs. Dibner (1991) argues that about 100 US DBFs founded during the last two decades no longer exist as independent entities, due to acquisitions, mergers and company failures.

DBFs' strategies have been formulated in the context of this turbulence. They emerged initially on the basis of the novelty of biotechnology. Their strategies have to a considerable extent relied on the concern of large firms to 'keep a handle' on biotechnology in such uncertain circumstances, by the use of collaboration. And DBFs have had to develop strategies to enable them to adapt quickly and flexibly as the uncertainties in the technology unfolded, and opportunities became delineated (Dodgson 1991c). As they begin to develop products the linkages they have with large firms remain important and occur in marketing and licensing agreements. An example of such collaborations, a joint venture between a leading US DBF, Amgen, and a Japanese brewer, Kirin, in one of the most successful biotechnology products, erythropoietin, is described in OTA (1991).

Kirin put up $12 million and Amgen contributed patent rights, technology and....
$4 million in its own funding. Research took place on both sides of the Pacific,

and the companies divided up worldwide marketing rights as follows: Amgen kept US rights, Kirin took Japanese rights, and the Kirin-Amgen joint venture itself held onto rights for the rest of the world. Johnson and Johnson later bargained for European marketing rights from Kirin-Amgen as well as rights to certain US markets from Amgen.

(OTA 1991: 61)

Some of the older DBFs have grown into sizeable companies, and they are themselves beginning to collaborate with smaller, start-up DBFs. Nearly one-half of all USA collaborations in pharmaceuticals occurred between DBFs in 1991 (Spalding 1991). So-called top-tier DBFs have now reached a stage where they can partner early-stage DBFs, and the reasons they do so are similar to the earlier large/small firm collaborations: namely to access emerging technologies and the creativity found in smaller organizations. The early-stage DBFs prefer to link with the established DBFs as they have similar corporate cultures, and understand their specific needs (Spalding 1991).

THE INSTITUTIONAL RESPONSE: COLLABORATIVE POLICIES

In many countries the problems of the evolution of biotechnology – the fact that it is science-based, the interdisciplinary nature of the science, and the difficulties of transfer into industry – have encouraged governments to play active roles in promoting collaboration in biotechnology. (Governments have, of course, also played important roles in the creation of regulations and intellectual property protection.) In Japan, for example, the Ministry of International Trade and Industry sponsors two collaborative research programmes: the Japan BioIndustry Association, which has 320 industrial companies, and the Research Association for Biotechnology. Through the Japan Key Technology Center a Protein Engineering Research Institute has been established, the functions of which are described in Chapter 10. In the Netherlands the government supports the Innovation Oriented Programme for Biotechnology to stimulate multidisciplinary cooperative research between the country's five university-based biotechnology centres, and the Industrial Stimulation Scheme to support high risk ventures in biotechnology and to promote technology transfer from academia to industry (OTA 1991).

Even in the USA, where the development of biotechnology has been much more freely encouraged by private sector initiatives, public policies designed to create research consortia, commonly at a state or local level, have played a role in the technology's development. An example would be the Midwest Plant Biotechnology Consortium which involves fifteen universities, three Federal laboratories and forty companies. While there are high numbers of university or research laboratory-based consortia, such multi-laboratory and multi-state consortia in the USA are unusual (OTA 1991).

In the UK, public policies towards biotechnology have included strong elements of collaboration promotion. These can be seen in the activities of two of the major bodies developed to encourage biotechnology: the Biotechnology Unit at the Department of Trade and Industry (DTI), and the Biotechnology Directorate of the Science and Engineering Research Council (SERC).

The Biotechnology Unit was formed in 1982. It places great emphasis on promoting the government's LINK programme which encourages industry to undertake joint research with universities and research and technology organizations. There are a number of LINK projects in biotechnology such as the Eukaryotic Gene Manipulation Programme. The DTI has programmes which support collaborative projects between companies in areas such as plant genetic engineering, advanced bioreactors and the bioseparation process. The collaborative, 'club' approach of the Biotechnology Unit has a number of manifestations, ranging from support for groups undertaking joint research to those just exchanging existing information. The Biosep club, for example, which has over 50 industrial members, offers regular up-to-date reports on bioseparations technology, and allows members to influence the programme of collaborative research being undertaken at a number of universities, and to have priority access to research findings. On the other hand, the Biotransformations club has just six members and concentrates on sharing the results of university research.

According to the official responsible for biotechnology clubs, this collaborative approach has been popular with firms and has a number of advantages, including:

1   large firms having the opportunity to observe what is going on in small firms, and small firms gaining access to large firm technology;
2   preparation for European integration, and the attraction of European partners;
3   awareness raising and linkage building;
4   improved industry–academia liaison;
5   shared risk in club research;
6   possibilities for commercial linkages outside of club activities (Rothwell, Dodgson and Lowe 1989).

The Biotechnology Directorate was established by the SERC in 1981. Its formation was a response to fear that Britain was falling behind in biotechnology research and that its interdisciplinary nature did not rest easily in the rigidly departmental structures of academia (which made industrial involvement that much more complex). It was also a means by which the SERC demarcated its activities from those of other research councils, such as the Medical Research Council, working in biotechnology.

The Directorate's objectives were to promote long-term research in biotechnology, support priority research sectors in order to develop centres of expertise that industry could tap, provide common resources in areas such as computing, and support postgraduate students. It has two other objectives: 'to develop collaborative activities with industry across a wide front including research', and 'the coordination of research programmes supported in the academic sector, linking academic groups with each other and with industry and proving a focal point for information about ongoing research programmes, including collaborative research'.

Activities in support of the latter two objectives include the extensive use of cooperative awards for projects involving firms and academic departments and the development of clubs and coordinated programmes of research. The club approach adopted by the Directorate is different from that of the DTI as it is more actively coordinated by an internal manager, and projects are peer-reviewed. Clubs promoted by the Directorate include: Protein Engineering, Antibiotics and rDNA and Animal Cell. These are usually 50 per cent funded by industry.

In an evaluation of the Biotechnology Directorate, Senker and Sharp (1988) argued that it had fulfilled its function well, and had attracted industrial sponsorship and involvement into well-defined research projects. It contended that

> The Directorate's methods are making an important contribution to the learning process of collaboration. For both industry and academe involvement in collaborative programmes of precompetitive research is new. It is a matter of building up trust and learning how best to work across disciplines and across institutions. Such learning only comes with experience and the Directorate's programmes have provided valuable experience for the British firms in biotechnology.
>
> (Senker and Sharp 1988: ii)

It reported, however, that there was some justification in the criticism of the club approach in the way it excluded smaller firms.

The other major public-sector player in biotechnology in Britain is the Medical Research Council (MRC). It has played a critical role in the development of the science of biotechnology, and through promotion of collaboration, its movement into industry as well. One of the major mechanisms it used in building up industrial biotechnology was its close collaboration with Celltech, a company it was involved in establishing (Celltech will be discussed below). Furthermore, in 1986 it created a MRC Collaboration Centre, with some forty staff, to liaise with industry. For a discussion of the relationship between the MRC and Celltech, see Dodgson (1990).

A STUDY OF COLLABORATION IN BIOTECHNOLOGY: CELLTECH/AMERICAN CYANAMID

Created in 1979, Celltech is a 350-employee DBF, located in Slough, England. In 1991 it had an annual turnover of £17 million. It is highly R&D intensive; in 1991 it spent over £10 million on R&D (such high expenditure is not unusual in DBFs).

In 1986 Celltech began to work under contract with American Cyanamid, a large USA drugs company. This contractual arrangement was revised into a cooperation agreement between the two companies in 1990.

Celltech is a science-based firm, and its necessarily high commitment to R&D in a developing business area has meant that the opportunity for making profits has been limited. It has recorded profits in only one year of its existence. It experienced considerable trauma in 1990 when its major shareholder went bust, and the company has subsequently experienced extensive operating difficulties. Contract R&D has given the company the chance to capitalize profitably on its considerable scientific capabilities. The contracts undertaken with Cyanamid have been the largest Celltech has undertaken and had brought £10 million into the firm by 1990.

Under the early contracts, in 1986, 1988 and 1989, Celltech undertook the R&D and the intellectual property rights (IPRs) were transferred to Cyanamid. Celltech's services were rewarded with a profit, but no exploitation rights (although Celltech has, as we shall see below, obtained many benefits from this arrangement). Under the new cooperative agreement, Celltech and Cyanamid share the scientific input. If the R&D proves successful, then Celltech will retain valuable IPRs. This could give Celltech its first experience of marketing products, and will enable it more fully to share the revenues from its discoveries. The evolution of this partnership reflects a growing confidence and trust between the two firms.

Celltech is a world leader in a highly advanced scientific field of MAbs. It has considerable expertise in the R&D of these antibodies, and in their manufacture. Particularly advanced is its skill in 'engineering' these antibodies to make them function more efficiently. The focus of the arrangements between Celltech and Cyanamid has been the use of MAbs to target cancers. Grossly oversimplifying, the monoclonal antibody is designed to target a specific tumour, and to deliver a means of destroying it. Two broad methods of destroying the cancer have been experimented with: radio-isotopes and toxins. Celltech possesses skills in the design and manufacture of monoclonal antibodies which Cyanamid does not. Cyanamid possesses a great deal of skill and knowledge in the field of toxins which Celltech does not. In this respect the two companies' scientific and technological know-how and capabilities are complementary.

Celltech has very considerable experience of working collaboratively. It has always enjoyed very extensive and strong links with academic researchers (see Rothwell and Dodgson 1991), and it has worked on a contract basis for many of the world's major drugs companies. It is currently working collaboratively with Merck, Sharpe and Dohme, the world's leading drugs company, Ortho Diagnostics, a subsidiary of Johnson and Johnson, and Genentech. Celltech also has experience of forming joint ventures. It has twice in the past formed joint ventures with Boots Company and Air Products. Neither of these joint ventures proved commercially successful, and both have been wound up.

The origins of the first deal with Cyanamid lie with a research package of Celltech's capabilities put together by its scientific directors and some marketing personnel. This package, involving the R&D and manufacture of monoclonal antibodies for the diagnosis and treatment of cancer, was developed in 1985. Around ten companies were approached with this package. The approach used by Celltech was novel; it was a case of telling firms 'we have good science, a good idea, fund us to find out if it works'. This novelty, and some scepticism about the scientific feasibility of the project, restricted to three the number of firms seriously interested in Celltech's proposal. Of these Cyanamid was favoured by Celltech, as it had considerable interests in the cancer field, and had been working with MAbs (although it had yet to build any capability in the area). The actual period of negotiation between Celltech and Cyanamid's UK and USA operations took around six months.

The focus of the arrangements between Celltech and Cyanamid is scientific research, and outcomes from it are inevitably long-term. There has, in consequence, been little tangible output from the relationship in terms of products (especially as the development time of new drugs is very lengthy due to regulatory requirements). However, a number of potential products are in the pipeline, including two which are planned to enter clinical trials in 1992. Scientifically, the research has proved productive. It has produced a family of patent applications in three important scientific areas, and by 1990 had resulted in around thirty patent applications being made. In two of these areas, where the results are not deemed confidential, around a dozen academic publications have been produced, based on the research.

Celltech has benefited in a number of ways from the arrangement with Cyanamid, apart from financially. Within Celltech three generic types of technology are conceptualized: background, enabling and project specific. These are schematically displayed in Figure 5.2. The classification relates to the interactions between research which can be done for contract clients and the company's own-products and technological development. Background technology is all the technology existing in the company at any point in time which is not contractually allocated in some way. In contract terms, it is pre-existing technology belonging to Celltech before the start date of a new contract. When undertaking any contract, both project specific and enabling technology is generated. Project specific technology is directed towards, and is applicable only to, the objective of the project. It is very tightly defined, and it will belong, or be licensed exclusively to, the project sponsor. Enabling technology is developed during the contract, but is applicable in wider fields than that which is the objective of the project. Celltech retains this technology but it would license it non-exclusively to the project sponsors insofar as it was necessary to practise the project specific technology. It is Celltech's strategy when undertaking contract research to build up enabling technology so that in subsequent projects it becomes background technology. In the programme of research with Cyanamid, Celltech retained all rights to non-cancer discoveries. In this way Celltech built its enabling technology and added to its broad technological skills and competences.

Since the reformulation of the arrangement with Cyanamid, Celltech has taken on much more of the risk of the development of potential products (thereby increasing potential returns). It is therefore funding the project-specific technology in a number of areas.

From the start of the arrangement with Cyanamid, Celltech had a project manager responsible for it. A top British scientist was recruited especially to run the initial contracts. As the programme was scientific in nature it was considered that a scientist rather than a professional manager would be the best person to manage it. However, considerable efforts were made to develop the management skills of the scientist. The project became the responsibility of the company's Director of Biology in 1990.

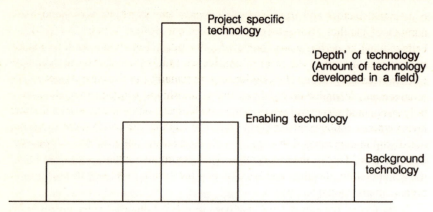

Figure 5.2   Conceptualization of technology

The scientists engaged on the collaboration are organized in a highly flexible manner to enhance creativity. According to the original project manager, it was Celltech's speed in development times that initially proved attractive to Cyanamid.

Apart from the managerial tasks of dealing with budgets and deadlines, a major part of the project leader's job is involved with communication. Keeping the partner informed is seen as crucially important. Ensuring that suitable internal communication occurs from the project to senior managers is also a priority. The series of contracts with Cyanamid have not been entirely bilateral. A national research laboratory, a university and a hospital have played important roles in the conduct and outcomes of the research. The management skills required have also included those of operating a research network of this nature.

One of the major problems that has been experienced in the collaboration, from Celltech's perspective, has been the establishment of a formal communications line into Cyanamid. Cyanamid is a vast, complex organization, and it took Celltech a number of years before it understood where executive power lay, and with whom it had to deal to get decisions made. Unlike Celltech, Cyanamid did not have a project manager, and in the early years of the arrangement dealt with someone from the technology side of things, without access to someone from the commercial function. This problem has now been sorted out, but it proved very difficult for Celltech.

Another potentially disruptive set of circumstances occurred in 1988 with the retirement from Cyanamid of two senior scientific managers. These managers were highly supportive of the deal with Celltech, and their replacements began to review all such arrangements. Initially sceptical of the scientific objectives of the programme, these new managers came to support them. This highlights the importance of personal factors and trust in the conduct of collaboration. Cyanamid has a number of parallel arrangements with other companies. Within Celltech it is believed that its performance compared to the other firms with which Cyanamid collaborated increased the level of confidence in it. Much of the trust and confidence necessary within the relationship, particularly in the early stages of the arrangement, depended on good personal relationships. Celltech enjoyed particularly strong and respectful relationships at chief executive and scientific director levels with its counterparts in Cyanamid. The fact that the collaboration survived this disruption to personal linkages reveals the existence of a level of inter-organizational trust. That is, the organizational culture, routines and procedures support the collaboration, the aims and benefits of which are well-known throughout the organizations (Dodgson 1992a).

A number of reasons are suggested by Celltech's managers and scientists for the success of the arrangement with Cyanamid. These are, in no particular order:

(a) The complementary nature of the partners' expertise and the mutual respect for each others scientific capabilities. Both partners possess equivalent levels of professional skill.

(b) Widespread acceptance within both partners of the need for collaboration and of the benefits from it. Each party understood the aims, objectives and goals fully as these were discussed at length before the programme began. To some extent this is a result of fortunate timing. At the time when the programme was first raised at Cyanamid, Celltech personnel felt that they had approached the company at the right time. They felt that biotechnology, and new biotechnology firms, had a high profile, and that engineered MAbs were very attractive to Cyanamid.

(c) Very considerable communications efforts, with Celltech attempting to inform Cyanamid of developments quickly and honestly.

(d) Good project management skills. Celltech is experienced at project management, although the programme of research with Cyanamid is the largest it has undertaken. The efforts made to recruit and train the right person to manage the project have been critical to its success. The continuity of project manager from the start of the programme has also assisted its successful progress.

(e) The development of trust and confidence. Cyanamid had a number of parallel arrangements with other companies. Within Celltech it is believed that its performance compared to the other firms with which Cyanamid collaborated increased the level of confidence in it. Much of this trust and confidence, particularly in the early stages of the arrangement, depended on good personal relationships.

CONCLUSIONS

The example of the development of biotechnology shows the extent of inter-organizational collaboration in the innovation process. Linkages with the science base –

as the source of new technological capability – and between established and new, small firms, with their particular advantages in state-of-the-art technology and production techniques – are very common in biotechnology and have played a significant role in its commercialization. These linkages have focussed on the acquisition of complementary capabilities. Large, established firms needed to access a new technology very new to them, and did so by linking with the DBFs that had built up significant scientific capabilities in the area. This was seen in the case study of a biotechnology collaboration. Some of the management problems of collaboration were described: partner selection, communication paths, project management and the building of trust to enable flexible and adaptable agreements. Celltech and Cyanamid enjoyed particularly strong and mutually respectful inter-personal relationships. When these were disrupted, the collaboration continued, as it had become engrained in the companies' routines and practices. They enjoyed a high level of inter-organizational trust.

Chapter 6

Collaboration and innovation – networks, intermediaries and standards

Innovation, it was argued in the last chapter, is a complex and iterative process involving a multiplicity of inputs. In the development of biotechnology, the science base, large and small firms have interacted together, sometimes with the assistance of public policy support. This chapter argues that there are a number of factors which complicate even further the relationships between all these actors. In the following discussion of three essentially separate issues: networks, intermediaries and standards creation, some of the complexity is revealed of relationships which have both competitive *and* cooperative elements, and involve intermediaries and encompass *political* considerations and negotiation.

REGIONAL NETWORKS

In Chapter 4 reference was made to various theoretical approaches to networks: dynamic networks, innovation networks, and the phenomenon of 'clustering' or regional industrial networks. All these approaches assume that networks – literally, interwoven organizations – possess advantages for competition and/or innovation. As pointed out, the concept is extremely broad. Here the position is taken that:

(a) the case of networks being *the* organizational form for competitivess in the 1980s and 1990s (Miles and Snow, 1986) is vague and unproven;
(b) theories of 'flexible specialization' of sustainable competitiveness based on closely linked and mutually supportive small firms are empirically contestable (see, for example, Kelley's (1992) criticisms of the assumptions of the suitability of flexible manufacturing equipment for smaller firms; and Cooke and Morgan's (1991) work describing the continuing centrality of large firms within regions noted for small firm vigour (also argued by Storper and Harrison 1991).

However, the role of frequent and intricate inter-organizational linkages in the process of innovation is well known (see the discussion in Chapter 2), and these are often regionally based. Lundvall (1988) discusses the problems of information exchange across geographical and cultural space. He argues that in the absence of accepted standards and codes able to transmit information, such as occurs when new technologies are developing and diffusing, information exchange depends on face-to-face contact and a common cultural background.

> When... technology is complex and ever changing, a short distance might be important for the competitiveness of both users and producers. Here, the information codes must be flexible and complex, and a common cultural background might be important to establish tacit codes of conduct and to facilitate the decoding of complex messages exchanged.
>
> (Lundvall 1988: 355)

Networks in the sense used here are, therefore, a convenient term for the complicated range of inter-relationships between industrial firms and the public and financial institutions which have historically proven to be so important in the creation and maintenance of innovative firms, and so often have a regional focus. They are also argued to have economic advantages: 'Regional economies displaying dense inter-firm and public–private interactions of this kind may be expected to show better than average growth performance' (Cooke and Morgan 1991: 24).

It is argued that small firms benefit particularly from networking. In their study of the West German foundry industry, in which the vast majority of firms are very small, Bessant and Grunt (1985) found that the skills and technological advantages that these firms enjoyed derived from their industry association and their preparedness to share licences within a wide network of firms. Much of the evidence to suggest that networks have proven efficient mechanisms for small company innovation and dynamism in certain regions comes from the 'Third Italy' (Piore and Sabel 1984).

Networks are argued to possess distinct advantages in innovation compared to large, vertically integrated firms. These are summarized by Hobday (1991) in his review of the literature on networking. Some of the advantages are claimed to be:

1 Groups of firms are able to remain at the leading edge of the technology via specialization and skill development.
2 The network can generate a source of constant innovation across the product, process and design spectrum.
3 Regionally-based institutions (e.g. consultancy firms, trade associations and financial backers) can support and provide valuable information to firms in the network.
4 Skill accumulation and collective learning occurs within the network, supported by the various institutions.
5 The network promotes flows of key individuals between firms, enabling them to develop and exploit their talents.
6 Skills can be combined and recombined to overcome bottlenecks and supply new innovations.
7 The concentrated focus of the innovative firm can reduce both the cost and time of a new innovation.

8 The network provides an entry route into the industry for small innovative firms.
9 Small firms are forced to find new low cost ways of developing new designs, new approaches to problems, etc.
10 Flexibility and low overheads enable networked firms to perform tasks large firms could only do relatively slowly and expensively.
11 Key individuals are attracted to dynamic new firms and the personal rewards for innovation can be extremely high.

Hobday finds an element of truth in these advantages in his analysis of the semiconductor industry. He sees, for example, that networked firms have considerably more flexibility in responding to technological opportunities than do vertically integrated and multi-divisional firms. Networked firms are directly linked to the market, and do not produce for internal corporate markets as do many vertically integrated and multi-divisional firms. They therefore have this advantage in their innovative activities. The fluidity of skilled personnel, both managerial and technological, combining and recombining in different firms greatly stimulates innovation. And firms combining to produce for niche markets can stimulate innovation in products and cost reductions. However, Hobday remains sceptical about the sustainable competitive efficiency of network organization compared with large firm organization, and finds that many of the benefits of networked firms are exaggerated.

He argues that individual firms in the network suffer from:

(a) Limits to growth imposed by shortages of finance and assets.
(b) Limited access to global marketing outlets.
(c) Being confined to locally generated, regional niche markets.
(d) A regional and inward-looking approach.
(e) Inability to adequately address international market and technological needs.
(f) Lack of in-house, large-scale manufacturing facilities often necessary to exploit innovations in the market place.

The problem with much of the analysis of regional networks starts when attempts are made to move beyond the 'cluster' type approach of the benefits of close inter-firm relationships for innovation and company growth, to those which portray it as some form of alternative across all industries to the previous logic of large-scale industry (Piore and Sabel 1984). As Bianchi and Bellini (1991) point out, it is only 'under very specific historic conditions (that) an organization of production based on small companies linked by non-hierarchical relations could be competitive' (Bianchi and Bellini 1991: 488).

As pointed out by these authors, and by Cooke and Morgan (1991), such circumstances depend on the decentralizing and rationalizing strategies of large firms, and that in some of these 'industrial districts' large firm market predominance is being regained. So, for example, in the dynamic Baden-Württemberg region of Germany:

It is the combination of large firms such as Daimler–Benz, Porsche, Robert Bosch, Audi, IBM, Hewlett–Packard and Sony with the large and dynamic SME (small and medium-sized enterprise) population that helps give Baden-Württemberg its edge in the contemporary economic order.

(Cooke and Morgan 1991: 26).

Although Porter's (1990) 'cluster' analysis of regional and national competitive advantage places large firms centrally within the network, his reliance on competitive rivalry as the means of building and sustaining economic growth fails to account for the valuable insights of some of the other network analyses which emphasize cooperativeness within them. Accounts of industrially and technologically successful regions and districts such as Silicon Valley and Route 128 (Roberts 1991), and Baden-Württemberg and Emilia-Romagna (Cooke and Morgan 1991) describe self-supporting cultures with high levels of trust between firms. These features, which are important for regional success, cannot simply be analysed by means of competitive market 'rules'.

In practice, the operation of networks requires both competition and cooperativeness, and it is this which makes them so difficult to manage. It also becomes more complicated when examining the innovation process in networks, as interlinked firms may be operating cooperatively with some firms in the process, and competitively with others. In other areas, the relationships may be reversed so that cooperating firms are competing and vice versa. The innovative advantages of networks, such as employee mobility and market focus may, as Hobday (1991) and Lundvall (1988) suggest, only be a temporally specific feature of industries affected by early-stage technological change.

CONTRACT RESEARCH ORGANIZATIONS

In their search for help in innovating, firms have a wider choice in collaborating organizations than just higher education insitutions, government laboratories and other firms. They can also work collaboratively with Contract Research Organizations (CROs). These organizations perform research on a membership or contract basis. There are a significant number of CROs in the UK, France and Germany, and these are important contributors to national R&D efforts (Bossard 1989). In Britain there are around 70 CROs, undertaking research valued at £670 million in 1988–89 (Ringe 1991). The 45 British CROs which are members of the Association of Independent Research and Technology Organizations (AIRTO) employ over 8500 (Rothwell, Dodgson and Lowe 1989).

CROs provide some of the oldest examples of formal collective industrial research. The British Research Associations (RAs), for example, were formed in the 1920s. According to Freeman (1991),

They were seen as a means of sharing the costs of acquiring technical information and of testing facilities, pilot plant and prototype development. They were thought to be mainly a device for overcoming market failure in industries where the threshold costs of R&D and other scientific and technical services were too high for small firms.

(Freeman 1991: 501)

Despite the belief that RAs would primarily provide a source of R&D for firms without in-house capabilities, Freeman cites a 1961 Federation of British Industries' survey which showed that RAs were actually used intensively by firms who had their own R&D. The RAs were thus 'an important ancillary and *complementary* source of scientific and technical information rather than a *substitute* for indigenous innovative activity' (Freeman 1991: 501). This model has, as we shall see in Chapter 10, been repeated in Japan.

Contemporary CROs (which include the RAs in new guise) offer a variety of services including contract research on a one-to-one or group basis, consultancy and testing. The role that these organizations play in promoting innovation is argued to be changing.

> During the 1960s, with new technologies emerging, a different form of CRO came to prominence. The focus of these organizations was the use of new technology and developing expertise in technology rather than particular industrial sectors. They marketed themselves as technology driven organizations able to improve customer's productivity through the introduction of new and appropriate technology, and also through reviewing, assessing and updating a customer's product design, marketing, processing and overall business planning. These organizations worked very much as equals to their customers – a customer brought in the CRO not to solve a particular problem in a prescribed way (although this was, and still is, one of the introductions a customer may have to a CRO), but to secure an informed analysis of the problem and to exploit the expertise and experience of the CRO in finding solutions, possibly in unexpected ways or areas. ... CROs are, in many cases, in a good position to assess a customer's technological capabilities in the light of both emerging technologies and of his general position within the marketplace in relation to competitors and the general industrial market, and then to follow up such assessments by introducing/developing any required technology.

(Ringe 1991: 19)

In his survey of 138 British industrial users of CRO services, Ringe (1991) found that firms were making increasing use of contract R&D as a way of deploying their R&D resources more efficiently. Firms saw using CROs as a cost effective way of accessing technological expertise they did not possess. They tended to perform more of their 'precompetitive' research on a collaborative basis in order to be involved in a new research area and to watch what their competitors were up to. Research 'closer to the market' tended to be contracted to CROs if the firms themselves could not do it (this commonly occurs with smaller firms without in-house R&D capabilities).

The increasing closeness between CROs and their clients is also described by Haour (1991):

> It is a bilateral, formal linkage, limited in time, targeted at a specific objective, and requiring, prior and during the collaboration, considerable intimacy and mutual understanding between the two partners with regard to expectations, goals and capabilities: for the duration of the project, the contractor constitutes a true extension of the client's R&D organisation.
>
> (Haour 1991: 2)

The role of these intermediary type organizations can also be proactive. In their study of the German foundry industry, Appleby and Bessant (1987) found that the foundry technical association played an important role in articulating the needs of small firms, so that research was directed and funded in line with the requirements of small firms in ways the small firms themselves could not elucidate, and occasionally did not recognize.

An example of the sort of role CROs perform is provided by CERAM Research, a 70-year-old, £5 million turnover CRO working in the ceramics and related materials field. Its business aims are to enhance the competitiveness of its customers through improving their products and processes through R&D, consultancy and training. In 1990 it had over 250 member companies, forty of which are from overseas.

The changing nature of the funding and functions of CERAM Research is typical of many CROs. In 1978, 67 per cent of income was derived from income from members, collaboratively sponsoring research; the remainder came from contract income. By 1990 this had declined to 32 per cent, and contract income increased to 50 per cent (the remainder came from government grants). CERAM Research's change in emphasis towards more contract research, i.e. moving closer to specific commercial applications, is mirrored by other AIRTO members getting research 'closer to the market'.

Examples of CERAM Research's collaborative research are a three-year programme on leadless glazes and decoration for its members, and work on emission monitoring in the heavy clay industry. The work on environmental services provides a good example of technological activities in the fields of regulation and monitoring which many individual companies cannot undertake, but can in collaboration. There is increased collaboration between CROs, and CERAM Research looks for technology in other industries which can be adopted in the ceramics industry. For example, in 1987 CERAM Research won a Queen's Award for Technological Achievement for a printing system for the decoration of tableware. This involved accessing technology from the rubber, printing and photographic industries. On the manufacturing front, CERAM Research and SATRA (the footwear CRO) jointly developed a CADCAM package. In these ways CROs can scan other industries and technologies in ways individual companies cannot. CERAM Research and the other CROs also play a role in technology transfer from higher education institutes.

STANDARDS

Technical standards can be established by standards authorities, such as the International Organization for Standardization (ISO), by voluntary agreement within an industry, or may exist *de facto* in line with the standards of predominant companies. There has been a rapid growth in the number of technical standards. Reddy (1987) argues that in the previous ten years, more international standards were generated than in the prior thirty years. The vast majority of standards are not *de facto*, and involve negotiation between companies, academics, standards authorities and other government departments, and this provides an important forum for collaborative activity.

> Industry-wide technical product standards play an important role in creating and mediating technological interdependence in industrial markets. Standardization activity is the domain of common interest and shared knowledge among producers of the same product, producers of complementary products, and the users.
> (Reddy, Cort and Lambert 1989: 14)

Standards are argued to be very important for the development of a new technology. Reddy (1990) provides examples of the way work on Integrated Services Digital Network (ISDN) demands cooperative action on the part of phone companies, satellite suppliers, microwave vendors, local area network companies, and value-added network operators; and the way Computer Integrated Manufacturing (CIM) involves linkages between mainframe and minicomputer manufacturers and suppliers, turnkey CAD systems, graphics workstations, production systems, unbundled software, systems integration services and other related hardware. The problems in developing common standards in the latter are argued to have delayed higher levels of system integration. David (1986) argues the importance of standards in assisting the diffusion of technology.

Standards are argued to have a wide range of benefits, summarised as:

1   reduction of transaction costs, by improving recognition of technical characteristics and avoidance of buyer dissatisfaction;
2   provision of physical economies by simplified design, production economies and ease of service;
3   advantages to buyers through interchangeability of suppliers, better second-hand markets and spare parts suppliers, and enhanced competition for sellers;
4   increased product innovation (except in the case of *de facto* standards) (Reddy, Cort and Lambert 1989: 18).

Collaboration R&D programmes can play an important role in the development of technical standards. As Mytelka (1991) argues

> firms in dynamic knowledge-intensive industries see the need to develop standards at a far earlier stage in the production process. Collaboration in precompetitive R&D of the sort promoted through ESPRIT has a particular advantage in this respect because it ensures an early approach to the harmon-

ization of technical solutions – environments, architectures, interfaces. At a later stage in the investment process, common standards enable firms to develop compatible products and this, it is argued, is a prerequisite to the creation of new markets in these dynamic sectors.

(Mytelka 1991: 197)

Indeed, Reddy, Cort and Lambert (1989) argue that the nature of standardization activity varies across stages in technology life cycles. Thus, activity in the emerging stages of a technology focusses on the creation of a common language, 'nomenclature and symbols'. The early stages also begin to address performance expectations, inspection, testing and certification procedures. As 'dominant designs' in the new technology emerge, standards focus on dimensional and variety reduction. The process of standard formation is never static and continues to be evaluated and revised throughout a product's life.

When technological development is very uncertain, and there is a wide range of conflicting approaches to technical and market problems, there are many advantages for companies participating in the standards-forming process.

Product standards, by their very nature, mediate and create interdependencies in industrial markets. In their role as mediators, products standards and process of standardization introduce an element of technological and operating stability into industrial markets.

(Reddy, Cort and Lambert 1989: 18)

Collaboration in the creation of standards provides a mechanism for reducing uncertainty and converging diverse technical efforts in firms. With the participation of numbers of companies, and experts from academia and government, the possibilities of producing 'objective' standards, rather than the 'subjective' standards of individual firms, and hence better technical sophistication, is much improved. Standards therefore possess both an industrial logic and a public good.

Before discussing an example of a collaboration formed to benefit from standard formation, it should be pointed out that standards also have an anti-competitive element. This is obviously the case in *de facto* standards, where control is exercised by single companies, but there are also examples of collusion on the part of firms and governments to provide exclusionary standards. Lamming (1992), for example, provides the example of Prometheus, the collaborative European 'intelligent highways' research programme, which is designed to develop a technical standard which will exclude competitors (non-European firms will have to comply with the standard, and this will take time and allow European firms an advantage). The exclusion of non-participants in the process of standards formation may explain why companies feel the pressure to collaborate so as to avoid the high costs of them not doing so.

As standards have competitive value, their formation is often complicated and controversial. Competing standards may exist, as seen clearly in the current case of High Definition Television. Parallel standards can exist side-by-side. Bessant

(1991) describes the way the German car industry at one time used five different standards. The evolution of standards can be a lengthy and highly competitive process. It is, furthermore, an extremely complex process. At present the standards-making bodies in information and telecommunications technologies include over ten major standards groups, and many hundreds of working level groups. The European telecommunications standards body, ETSI, for example, has 12 technical committees, 50 sub-technical committees and several hundred working parties. Its USA equivalent, TI, has over a thousand committees, and the ISO has over 7000 standards in information technology. The fact that many of these bodies have to work internationally introduces the question of politics. The political nature of standards creation is one of the features of the following case study.

CASE STUDY: GEC SENSORS – TID/ALCATEL

The Tactical Information Division (TID) is one of five autonomous businesses within GEC Sensors which is itself a division of the GEC–Marconi sub-group of the GEC conglomerate. TID has three principal spheres of business activity, communications, data systems and in-flight telephones. Its market is almost exclusively the military and civil aviation industries. TID employs around 100 people, the majority of whom are technologists and technicians. Annual turnover is in the region of £10–15 million. In order to win large contracts, the company has to engage in long periods of costly development work which can make revenues somewhat lumpy. The uncertainty affecting the international defence industries is not expected to affect TID in the same way as many other areas of GEC. TID is a systems house and its 'defensive' products will still be in demand even within an extensively rationalized military.

In 1990 a collaboration between GEC Sensors and the French company, Alcatel ATFH, was formalized through the Eureka initiative (see Chapter 7). Its aims are to demonstrate a terrestial flight telephone service (TFTS) for airborne public correspondence. GEC Sensors is the lead partner for the purpose of Eureka, but in practical terms the collaboration is between equals. Alcatel has three relevant communications buinesses: in radio and microwaves; networking; and telecoms switches and PABXs.

The TFTS project employs around forty-five TID personnel: five managers, ten research scientists and thirty engineers: this includes a number from other parts of GEC. Alcatel has a similar complement of personnel. The planned cost for the three-year project is 40 million ECU, although this will be revised in the light of findings from the development and test stages. The finance is predominantly supplied by the two partners. Alcatel, however, has the development costs under-written, in part, by both Air France and the French government. TID is primarily self-funded. It has committed several million pounds to the Eureka project: the development of the specification alone cost TID in excess of £1 million. It has received a nominal grant from the DTI, believed within the company to be perhaps ten per cent of the public support enjoyed by Alcatel.

The GEC Sensors–Alcatel TFTS is a totally digital, two-way communications system. There are three subsystems in TFTS: the aircraft subsystem (the responsibility of GEC Sensors); the ground station subsystem (Alcatel's responsibility); and the support subsystem (to be put together by interested telecoms operators). The aircraft subsystem is made up of leading-edge components to meet the size, weight and reliability requirements of such an hostile environment. GEC Sensors has drawn extensively on its military-derived technological know-how. With the advances in technological sophistication and quality now being demanded by commercial markets, TID is seeing more opportunities to use its defence-related competences.

The ground station subsystem is similarly novel. It involves a pan-European networked grid of fifty stations that provide a total communications coverage for air traffic across Europe. There are another 100 stations at major airports around Europe which provide connections for on-the-ground aircraft and aircraft during their ascent to their working altitude, whereupon the European grid takes over. Both of these subsystems have entailed the development of some highly advanced control and communications software, to control switching systems, air-ground communications, real-time customer billing, service provision and monitoring and maintenance systems.

According to TID's Divisional Director, it is capacity (current and projected) which drives and dictates developments in communications technologies. The collaborators were forced to underwrite the development of a terrestrial (microwave) communication system rather than use existing and proven satellite technologies because of the capacity and bandwidth limitations of the latter, and because of its high purchase and installation costs.

TID's rationale for initiating an international collaborative venture through Eureka was simply that it represented a political means to a commercial end. TID recognized that a pan-European TFTS has implications for the communications industries, communications service providers and the airlines of all European states. In order to bring about a commercially viable European TFTS, the development work and standard setting had to be pan-European.

TID's commercial reasoning was simply that it was well placed to develop a lucrative European communications market. Much of this related to the question of developing standards in the field. The European Telecoms Standards Institute (ETSI) was arguing that the emerging European Groupe Speciale Mobile (GSM) standards and technologies for land-based mobile communications were not suitable for air-to-land correspondence, and there was an obvious market for GEC Sensors. The TFTS project also provided an opportunity to support TID's wider efforts to reduce its exposure to the defence marketplace and, more importantly, to capitalize on its deep technological prowess gained through their activities on military contracts.

According to TID, Alcatel's interests were essentially commercial. The commercialization of TFTS was felt to represent a new, albeit limited, market for Alcatel's ground station and networking technologies. It was also anticipated that success in Europe was likely to be followed by sales in other regions such as Australasia and the Far East. Although not competing in this technology, GEC and Alcatel do compete in other product areas such as central switches.

TID had identified a need for a digital terrestrial flight telephone service some years ago. Its original interest was focussed on the UK and British Airway's domestic shuttle services. After some investigation it became apparent that the UK market and British Airways were not going to be able or willing to underwrite the costs that such a new service would entail. It increased its efforts in 1988, when the European Telecoms Standards Institute (ETSI) was formed and announced its interest in the development of a European-wide TFTS. ETSI's work on TFTS has been carried out through the RES5 technology committee.

The Europeans originally considered updating the US 'airfone' TFTS, but this system was based on analogue rather than digital technologies and was therefore inherently limited in its potential. Since it was agreed that TFTS should be conducted within the Eureka framework, the collaborators have spent much of their time meeting, discussing and lobbying the various and collective interests of the European communications industries.

This process of consultation has been critical to the project's getting off the ground. On paper the TFTS project is a two company, Anglo-French initiative attempting to set pan-European standards. However, a wide range of organizations have been involved. Senior management at TID and Alcatel have spoken with the major European airlines, and SAS, British Airways and Air France are assisting with the project. The partners have also approached communications systems manufacturers such as Ericsson and Siemens about the project. Siemens is considering active involvement in reviewing the definition of interfaces and protocols and the design of experimental stations.

TID and Alcatel have also contacted all of the European Post and Telecommunications Authorities (PTTs), and both British Telecom and France Cables et Radio are working on the TFTS project. Alcatel's relationship with the French PTT is believed to be much the closer of these relationships. TID and Alcatel have also approached the German PTT. The Bundespost has reservations about the viability of TFTS and has not wished to become directly involved. However, the Bundespost takes a leading role within the ETSI RES5 committee and was instrumental in the issue of a memorandum of understanding which sought to elicit the support of all European PTTs for the TFTS development efforts of TID and Alcatel. The Bundespost, it would seem, does not wish to be left out of this potentially important communciations innovation.

ETSI and TID's efforts to establish TFTS as an independent communications system with its own architecture and standards met with some opposition from the established GSM lobby. The consultation and lobbying process saw some different national positions being established. The Germans in particular voiced reservations about the commercial viability of TFTS and felt that it could dilute European communications industries' efforts to harmonize their much larger national land-orientated communications systems around the GSM standard. The French, on the other hand, were most vociferous in support of the TFTS initiative. There was also strong support from Scandinavian companies which were very active within ETSI and other European PTT forums in the promotion of TFTS and the Eureka project. ETSI has itself played an important role in promoting TFTS. ETSI has mandatory powers which, in part, gave TFTS the ability to combat pressures to concentrate on GSM.

GEC Sensors and TID were introduced to Alcatel initially by France Cables et Radio at an ETSI working party. TID managers suggest that the company eventually chose to work with Alcatel for a number of very specific reasons:

(a) they were not competitors in this area;
(b) they had complementary world-class capabilities in land-based communications, ground-stations and networking technologies. TID's expertise lay exclusively in aviation systems;
(c) TID had very limited experience of working with PTTs in particular and the civil sector generally. Alcatel had longstanding links with France Cables et Radio. The culture and working relationships of PTTs are very different from those found in military and civil aircraft;
(d) Alcatel was already a pan-European organization with business networks across Europe, whereas TID had been far more parochial in their activities. Alcatel's official language is English;
(e) Alcatel had a long history of successful non-military collaborative ventures including previous work with France Cables et Radio;
(f) Perhaps the most important factor was the rather more intangible feeling of both senior management teams that they could work together.

TID's Divisional Director indicated that the selection of a partner was a critical element in setting the scene for a successful collaborative venture. He argues that when companies consider such an alliance it is typical for them to have only a partial understanding about the particular sphere of activity, little or no awareness of suitors and probably limited appreciation of either national or organizational cultures. It is therefore difficult to be too prescriptive about how to set about selecting the appropriate partner. Conducting systematic profiles of sectors and organizations is costly, time-consuming and as likely to identify meaningless information as meaningful. GEC Sensors avoided such a rigid approach and the development of a selection checklist.

The selection process was not, however, left to chance; considerable time and effort was devoted to identifying the right partner. GEC Sensors used two senior managers with considerable collaborative experience within the European flight communications arena and relied on their intuition to identify a complementary organization with which it would be possible for GEC Sensors to produce a profitable TFTS. GEC Sensors' Managing Director and TID's Divisional Director both felt that Alcatel was by far the most suitable potential partner for the TFTS project, and they have collectively championed the initiation of the Eureka project.

Although the project is in its early stages, the collaboration from TID's perspective is showing promising signs. Both parties have expressed an interest in the interchange of production facilities. The hi-technology, development-intensive markets that both companies are exposed to are notoriously lumpy in terms of revenue generation and production workload. Within TID it is believed that collaborations can offer rapid access to the right kinds of production capacity without the debilitating cost of any one organization supporting that capability.

However, TID accepts that it takes time to develop such interoperability and that one of the biggest problems facing the partners is their very different orientation towards defence and civilian markets. TID would very likely over-engineer Alcatel's products, while Alcatel was thought likely to under-engineer TID's products. However, TID and Alcatel have agreed that, in the future, each would promote the other's equipment and there is a hope that the partners will be continuing in business of some sort for the next decade or more.

There is no contractual relationship between GEC Sensors and Alcatel. The latter's contract is with France Cables et Radio. There is no formal collaborative agreement detailing terms and conditions, simply a signed undertaking by senior managements at both companies to collaborate in the development of TFTS to their mutual benefit. Neither TID nor Alcatel have sought to impose too many management strictures on the project, preferring to leave the programme to evolve in a more organic manner. Indeed, Alcatel was concerned to manage the development stage of the project as a coordinated but essentially independent activity. In its experience too much proximity at this stage would lead to confusion and be counter-productive.

The coordination of activities and the process of knowledge transfer has been facilitated primarily through project meetings and the exchange of personnel. Representatives of the two project teams meet once a week to discuss progress and work through any operational problems. The TID project team also meet once a week in order to maintain momentum and direction. Project management and engineering change procedures will be formalized at the build and test stage, when both partners will need to evaluate their preliminary findings.

TID's Divisional Director indicated that communication across national and organizational boundaries was a central issue: both parties speak the same language but frequently mean and infer very different things. He argued that the solution in this case is the documentation of events and meetings and the writing of protocols to be agreed and then enacted by everyone. It is tedious and laborious in the short run, but in the longer term it can save many costly mistakes.

The apparent ease with which this loose-tight approach is being implemented was thought to be in part attributable to the extensive TFTS work carried out by TID prior to the Eureka project. GEC Sensors had to a large extent decided what the system should look like and made their commitment to the collaboration conditional upon Alcatel accepting much of this broad definition.

TID managers believe the most likely source of future complication involves the third area, the support subsystem. This will involve cooperation and support from PTTs for interfaces to the various European public switched networks;

cooperation from potential and actual TFTS service providers (not necessarily PTTs) in service provision, billing, maintenance, etc.; and all of this will be in part decided through a special *ad hoc* committee consisting mainly of ETSI RES5 members. Other problems may emerge from the entrenched interests of some of the PTTs which could well make the setting of standards for the architecture, interface and terms of reference for service provision, problematic.

Another potential area of difficulty which could also have resulted in project failure related to internal GEC management attitudes to project costs. Senior management in GEC Sensors believed that collaborative ventures were concerned with productivity gains and cost sharing: in short, they expected cost savings over earlier independent TID budgets. They are unlikely to materialize; indeed, within TID it is believed that collaboration is likely to cost more and take longer, and that its advantages lie with improved product scope. TID and GEC Sensors, it is argued, could have developed the technology within the GEC organization and certainly within the confines of the UK, but they could not have created a European market. With hindsight, the TID management also suggested that the collaboration with Alcatel would lead to a qualitatively better product. Indeed, it is thought that GEC Sensors would increasingly have to rely on collaborative initiatives to develop leading edge technologies for three principal reasons: the internationalization of activities, the growing sophistication and breadth of technology contained in products, and the sheer cost of underwriting such projects.

A number of factors are suggested within the company for the success of the collaboration:

(a) It addressed a common need: the companies had come together to address a number of clearly understood commercial needs; and the collaboration was meaningful and relevant to both parties.

(b) There was a complementarity in the two organizations' commercial and technological capabilities, but also in their philosophies towards collaboration. Put simply, TID management enjoyed working with their Alcatel counterparts.

(c) The apparently obvious division of responsibilities into air and ground systems had matched both partners' core technologies, and was believed to have contributed significantly to the ease and speed with which both parties began producing outputs.

(d) high levels of communication have been central to the ease with which both companies have been able to begin working together. Open communication has enabled Alcatel to quickly get up to speed in terms of understanding the machinations of a digital TFTS.

(e) Similarly, the extensive communication and consultation with other interested parties across the European communications industries has enabled GEC and Alcatel to get this Anglo–French collaboration off the ground as a pan-European initiative.

(f) Both parties have made a concerted effort to go beyond the strictures of language and culture and through intensive communication and protocols, arrive at a common understanding of events, needs, responsibilities, etc.

(g) senior managements at both companies, and at France Cables et Radio and Air France have been extremely determined in the setting up and framing of this collaboration. Project managers and operational staff have been equally determined and willing to make a success of the cooperation.

(h) timing was fortuitous: GEC Sensors was already thinking about a digital TFTS when the new regulatory body, ETSI, focussed European interest on such air–ground public communication systems. It was fortuitous that ETSI was formed at this time as the previous standards body, which was PTT dominated, would have been unlikely to sanction such potentially challenging developments in communications services.

(i) experience and good management have been essential. Senior management at both companies had extensive experience of collaboration, albeit very different kinds of alliances. They consciously allowed the possibility of collaboration to unfold rather than attempt to force it. Having initiated the alliance, they allowed the relationship to evolve organically according to need rather than imposing rigid and pre-ordained structures on it.

CONCLUSIONS

By using the distinct examples of networking, intermediaries and standards formation, this chapter has argued the extraordinary complexity of inter-organizational linkages in innovation. In the discussion of networking it was shown that close inter-firm linkages are characterized by both cooperation and competition. Firms compete in some areas and cooperate in others. For example, in the case study, GEC and Alcatel cooperated in TFTS, although they are competitors in other areas. In standards creation, there are many pressures towards jointly-established, unified standards. However, the cooperativeness needed to create standards exists alongside strong competition from firms knowing their commercial value. The growing role of CROs as important supplementary sources of R&D also adds to the complexity of the situation. Do firms contract out R&D to CROs whose knowledge will expand on the back of it and may be useful to competing firms also using the CRO?

These complexities lead to consideration of new forms of relationship. The CROs, for example, need to develop strong trust-based relationships with their clients to placate fears of giving away technology amongst contractor companies. The innovative advantages of small firms in networks (which are industrially and temporally specific) depend upon close and integrative relationships. Firms operating in political environments, such as standards bodies seen in the case study, need to learn new skills and develop new forms of relationship. GEC–TID, for example, was learning about collaboration as well as developing a new product. It was using the international political environment to form a new standard, and its external political lobbying and negotiation provided one important learning experience. Another was internal: the company was learning to deal with another company cooperatively in circumstances where previously it had only competed, and in an 'unfair' environment where its partner was receiving considerable financial support, and it was not.

The extent of these tensions and conflicts, which are endemic amongst collaborating firms, needs to be fully appraised so as to overcome the more naive assumptions of some of the networking literature. However, as this chapter has shown, these complex inter-organizational linkages can promote innovation. Future chapters will examine in greater depth the environments and forms of relationship most appropriate for achieving innovative advantage through collaboration.

Chapter 7

Public policy and technological collaboration

National and regional governments, and many pan-national organizations, such as the European Community, actively promote technological collaboration. The rationale and some of the forms of this promotion have been discussed in earlier chapters. The public institutional response to biotechnology has been described. This chapter provides a brief review of three public-policy programmes devised to promote collaboration: ESPRIT, Eureka and Alvey. Some of the positive and negative outcomes from these schemes are described. A case study of a firm's experience in the Alvey Programme is provided, and the complexity and continuing problems of participation analysed.

SUPPORT FOR COLLABORATION

Governments have used a variety of methods to encourage collaboration, including the formation of research associations and consortia, the relaxation of legislative restrictions, such as the removal of anti-trust laws for joint R&D in the USA, the creation of a variety of technology transfer organizations, and taxation policies. Examples of the latter are the tax exemption for R&D limited partnerships in the USA, which enabled tax allowances for joint R&D, and encouraged extensive partnerships in areas such as biotechnology; and the system in Japan that allows firms engaged in collaborative research to depreciate capital equipment 100 per cent in the first year of a project (Levy and Samuels 1991).

At the forefront of government policies to promote collaboration have been cooperative research programmes. A number of examples of these schemes are now discussed to illustrate some of the reasons for and outcomes of the promotion of collaboration.

THE ESPRIT PROGRAMME

In the late 1970s it became increasingly clear in Europe that European electronics firms were becoming uncompetitive internationally, and that the basis of this problem was the comparative technological advance of the Japanese. Unlike in Europe, the Japanese electronics industry had considerable experience of collaborative research efforts, and within the European Commission efforts began to replicate the Japanese example and to promote technological development through collaboration. The Japanese Fifth Generation Computer Programme in particular had raised fears in Europe (Arnold and Guy 1986). Discussions began in 1979 between the EC and the 'big twelve' European electronics firms on how to frame a collaborative programme. In 1982 a pilot phase project began to encourage collaboration between the major European companies and their smaller counterparts, and universities and research institutes.

This pilot project proved successful and the ESPRIT (European Strategic Programme for Research and Development in Information Technology) was launched in 1984 with the aim of fostering 'pre-competitive' R&D collaboration to provide standards and components to assist the international competitiveness of European IT firms. ESPRIT has operated in three stages. The first stage, ESPRIT I (1984–8), had a budget of £525 million and funded 220 contracts. ESPRIT II (1988–92) had a budget of £1.1 billion, and funded 369 projects. The third stage, a European IT programme (1990–4) was launched in 1991 with a budget of £946 million, covering microelectronics, information processing systems and software, advanced business and home systems and peripherals, computer integrated manufacturing and engineering, and basic research.

The basic level of financial support available to participants (firms, academic and research institutes) is 50 per cent of costs for firms, and 100 per cent for academic and research institutes. Projects must include industrial companies from at least two EC member countries. Two kinds of project are supported: large projects with already defined 'strategic' aims, and smaller more speculative projects operating in broadly defined areas. The latter category receives the highest proportion of ESPRIT funds.

In evaluating how successful ESPRIT has been, and its impact on European electronics companies, it is necessary to be conscious of its financial limitations. Although ESPRIT is the major European Commission technology initiative and has received a substantial budget (in the current Framework Programme, information and communications technologies account for around 40 per cent of total EC expenditure), its scale compared to that of some individual firms is limited. In the total period 1984–94 ESPRIT funds for R&D are roughly equivalent to the R&D expenditure of Siemens in 1990 alone. For some large companies working in numerous ESPRIT projects, such as the French company Bull, total ESPRIT project contributions have accounted for only 5 per cent of total R&D efforts (Mytelka 1991). Nevertheless, demand for ESPRIT funds has consistently and considerably outstripped their supply, and it has had a significant impact.

Mytelka (1991) argues that ESPRIT has had the following beneficial effects:

1   the sharing of risks and costs;
2   the range of applications for existing or new technologies has been extended by building inter-sectoral links;
3   improved observation of market trends and a reduction in uncertainty;

4   networking has promoted a 'critical mass in research and development by enabling larger European firms to secure complementary technological assets from a range of different project partners. This has enhanced the ability of these companies to respond flexibly to change without adding to the inertia of the firm' (Mytelka 1991: 191);
5   encouraged firms to establish European standards and harmonize interfaces 'in such a way as to overcome the disadvantages associated with the relatively smaller size of European firms as compared with their American or Japanese competitors' (Mytelka 1991: 196);
6   enhanced the knowledge accumulation possibilities of small and medium-sized firms.

In a study of French firm and research organization participation in ESPRIT projects, she also found:

(a) ESPRIT encouraged an international perspective;
(b) it improved the quality of R&D;
(c) it increased awareness of technological competences in other European organizations;
(d) it encouraged longer-term, more basic research in collaboration with higher education institutes.

Sharp (1989) is less sanguine in her analysis of ESPRIT. However, she argues that there have been some technological successes, and there have also been important *psychological* impacts. These include the provision of a channel for cooperation which previously did not exist; a mechanism for creating convergent expectations amongst top decision makers in industry; a constituency pressing for the completion of the European internal market with the removal of divergent standards and regulations; and an important learning process in collaboration. On the final point she contends that for many European firms ESPRIT has been the first experience of collaboration, and it has helped firms move up the learning curve. She further argues that policies towards collaboration may be a transitory phenomenon as industries restructure during a period of rapid technological change.

THE EUREKA PROGRAMME

Eureka is a pan-European collaborative technology programme which began in 1985 following an initiative by the French and, in some respects, as a response to the US Strategic Defense Initiative. It is both different and complementary to other European initiatives such as ESPRIT. Participants include the twelve EC countries, the six EFTA countries and Turkey. Unlike the EC programmes it is not directed into particular technologies or projects, and it is more concerned with 'market oriented' research. Its aims are to facilitate market oriented collaborative projects, led by industry, in all sectors of technology with the objective of introducing new products, processes and services.

Eureka projects can be in any technology, and the criteria for support are that the project involves technical innovation and includes organizations from at least two Eureka-supporting nations. The technologies which have received Eureka backing include communications, energy, environment, IT, lasers, medical/biotechnology, new materials, robotics and production automation and transport. Between 1985 and 1991 493 projects worth £5.9 billion have been supported (DTI 1991). Funding for the projects does not come from Eureka itself, but from national governments. Project proposals are circulated on a central data base to all national Eureka offices to allow the opportunity to find national partners before they are endorsed as Eureka projects.

Included in the Eureka initiative is the provision of networks to find partners, such as the EUROENVIRON project to find partners in environmental technologies. Projects have focussed on industrial, urban and agricultural wastes, air and noise pollution control, water quality, herbicide and pesticide pollution, environmental management systems, environmental catastrophes and clean production technologies. Another example of a network is the FAMOS project concerned with flexible manufacturing automated systems. In this project the aim is to develop systems which can be applied across a wide range of manufacturing industries from the electrical, automotive and aerospace industries to shoes, toys and construction. The FAMOS project has successfully developed a number of demonstrator plants.

The lack of strategic direction and aims of Eureka and its technological heterogeneity have made it difficult to evaluate. However, two questionnaire surveys of participants throw some light on the contribution of Eureka from the perspective of participants (Peterson 1990). Peterson found that:

1 Eureka projects tend to build upon already established links between organizations. Around two-thirds of projects leaders had previously collaborated with at least one of their partners.
2 Eureka extends the number of participants through circulation of details of projects.
3 Eureka projects tend to be rather mixed. Some are large and multi-partner, others involve only two or three partners. Most operate 'near market', but there are significant numbers of 'pre-competitive' projects.
4 The majority of projects receive public funding, and this tends to be between one-third and one-half of costs. It is broadly estimated that 35 per cent of the costs of all Eureka projects are met by public funds. While some firms have found Eureka status an aid to attracting private sector finance, governments have played little or no role in linking private and public funding.
5 Public funding is not the primary benefit for firms. Instead it is the benefits which accrue through collaboration, such as the cross-fertilization of ideas.
6 While up to one-half of Eureka projects would probably occur without Eureka support, these tend to be large firm-led projects with few partners, many of whom have worked together in the past. On the other hand there are projects with a degree of 'additionality', constituting up to around 40 per cent of the

total number, which might not exist without Eureka support. These tend to be led by small and medium-sized firms, RTOs and universities, and include partners inexperienced in collaboration.

7 Small and medium-sized firms faced problems due to their inexperience of collaboration, and the formal organizational and legal requirements.

THE ALVEY PROGRAMME

The Alvey Programme in advanced information technology ran in the UK from 1984 to 1990. Initiated as a response to declining performance of the UK IT sector, it was a collaborative programme aimed at improving industrial and academic science and technology, transferring technology from academia into industry, and enhancing competitiveness. The programme focussed on five main technologies: very large-scale integration (VLSI) semiconductor technology; intelligent knowledge-based systems; software engineering; man-machine interfacing; and systems architecture. The government provided £200 million to the programme, and industry approximately £150 million. Projects received 50 per cent of project costs from the government. It was managed by a Directorate, largely consisting of secondees.

Alvey supported 198 collaborative projects, with an average of 3.6 participating organizations, and lasting for an average of two to three years. Additionally, Alvey supported 113 'uncle' projects run in academic departments with industrial observers. Alvey involved 115 firms, 68 academic institutions and 27 government research establishments.

The Alvey programme enjoyed one of the most complete evaluations of a collaborative programme, the official evaluation running throughout the period of its existence (Guy and Georghiou 1991). The evaluation of Alvey examined the three sets of goals set by the programme. These were technological goals set by the projects' Directorate and participants; structural goals which addressed the research environment through improving academia-industry linkages and strengthening the UK's IT R&D base; and strategic goals to preserve and improve UK IT capability and competitiveness. The findings of the Alvey evaluation were positive. It is argued that the programme had:

(a) largely met the technological aims of individual projects. Industrial Alvey R&D tended to be in core technology areas, vital to their organization's interest and with generally long-term time horizons.

(b) consolidated and added to R&D capability. As Table 7.1 shows, it had allowed firms to develop new tools and techniques (this was also the most important response from academic participants). It had allowed firms to accelerate R&D activities beyond what was possible without Alvey support, and had upgraded skills.

(c) extended and improved linkages between an increasingly strong IT 'community'. However, the primary benefit occurred from academic participation in industrial research projects, rather than from improving industrial input into academic research, or partnerships within industry.

(d) proved a beneficial experience for participants (although many smaller firms found the costs high). The vast majority of participants had a positive attitude towards the programme and continue to build on Alvey projects, many with their Alvey partners.

(e) As for strategic aims, it is argued that it is still very early to see the commercial benefits of a 'pre-competitive' research programme, but there are some indications that these are occurring. However, the barriers to exploitation of research results are considerable, and are often related to internal management deficiencies, such as poor integration between functions and inadequate funding.

Table 7.1 shows that a high percentage (46 per cent) of Alvey participants regarded the programme as a means of conducting research which otherwise would

Table 7.1 What Alvey allowed industrial participants to do

Activity	Percentage of teams declaring Alvey to be very important		Achievement score*
Develop new tools and techniques	64	(1)	3.7
Accelerate R&D	57	(2)	3.8
Upgrade skills	52	(3)	3.9
Use new tools and techniques	51	(4)	3.9
Maintain R&D presence	46	(=5)	3.6
Access academic know-how	46	(=5)	3.2
Enter new R&D areas	45	(=7)	3.8
Build on R&D base	45	(=7)	3.8
Enhance image and reputation	38	(=9)	3.7
Establish new academic links	38	(=9)	3.5
Spread costs	37	(11)	3.6
Deepen understanding	35	(=12)	3.8
Develop prototypes	35	(=12)	3.5
Achieve critical mass	28	(14)	3.6
Keep track of R&D	27	(15)	3.6
Spread risks	25	(16)	3.8
Develop products	22	(=17)	3.8
Establish new industry links	22	(=17)	3.1
Access industry know-how	22	(=17)	3.1
Enter international R&D programmes	22	(=17)	3.3
Enter private sector R&D ventures	20	(21)	3.2
Enter national R&D programmes	15	(22)	3.2
Get used to new IT standards	10	(=23)	3.2
Influence new IT standards	10	(=23)	3.1
Enter new non-R&D collaborations	6	(25)	3.1

* Mean scores on a 1–5 'Achievements compared to Expectations' scale
Source: Guy and Georghiou 1991

have been cut. As the evaluation argues, this is 'disconcerting'. Research central to the core technologies of the UK's leading IT firms would not have been funded without government funding assistance. This highlights the major problem within British industry in funding technological development (House of Lords 1991). Other aspects of the Alvey programme highlight more of the strategic shortcomings of British firms, and how public policies can assist in overcoming these. Guy (1989), for example, argues:

> Alvey has impacted strategic and tactical thinking in a number of ways. For some the act of thinking through the position of Alvey work within the context of existing firm strategies has helped them translate tacit agreements as to the nature of these strategies into more explicit, articulated policies. Alvey has helped firms decide on the relevance of particular technology areas to their overall business stratagems. It has helped others rethink and restructure the way R&D fits into a firm. Furthermore... it has certainly caused almost all firms to rethink their attitudes to precompetitive and collaborative research programmes.
>
> (Guy 1989: 199)

An example of an Alvey project will now be provided to give some indication of the nature of such collaborations and the complexities and problems encountered.

CASE STUDY: RACAL ELECTRONICS

Racal is a large company which, when the Alvey programme began, included Racal Electronics as the parent company, with wholly-owned subsidiaries including Racal Milgo, Racal Redac, Racal Defence, Decca and Chubb. Racal's cellular telephone subsidiary Vodafone was floated as a separate company in 1989. The telecommunications joint venture Orbitel, which Racal jointly owned with Plessey, became fully owned by Racal following Plessey's takeover by GEC.

Racal Research Limited was founded in 1979 and employs around 200 people. Research activity is divided into eight areas: radio, digital signal processing, physical security, mathematics (security and cryptography), VLSI, CAD, software engineering and cellular. Racal Research supplies research across the whole of the Racal organization, but historically was most strongly linked to the radio division. Racal Research was involved in the original bid for the licence for the cellphone network.

Mobile Information Systems (MIS) was a collaborative R&D project funded by the Alvey programme. In addition to supporting 'pre-competitive' R&D in enabling technologies Alvey funded five 'Large Demonstrator' projects which were intended to provide exploitation-led goals for Alvey R&D, and to pull-through and demonstrate Alvey results as these emerged from the research programme. The Large Demonstrators, which involved larger collaborative groups than the Alvey 'enabling technology' projects, represented around 13 per cent of government investment in the programme.

MIS was a five-year project which started in September 1984 and cost a total of £7.5 million, including industrial investment. Membership of the participation changed somewhat during the course of the project, but the principal partners were:

Racal Research Limited   Cambridge University
Racal–Milgo Limited   HUSAT, Loughborough University
Hewlett–Packard Limited   Sussex University
Ferranti Electronics Limited   Thames Polytechnic
(later Plessey Semiconductors Limited)
The Electricity Council and   University College London
two Regional Electricity Boards

Racal Research was project manager for the MIS project. As with virtually all Alvey collaborative projects there was no central site, collaboration being conducted by linking distributed research at each partner's site by means of electronic mail and regular project meetings. Effective project management was thus essential for the success of the collaboration. A total of sixty to seventy researchers were involved in working on the MIS project, with a maximum of forty to forty-five at any one time.

The collaboration was initiated when the Alvey Directorate approached Racal senior management to invite the firm to put a project forward for a large demonstrator. Racal senior management asked Racal Research to formulate a project proposal. Racal Research had recognized prior to the Alvey Programme that to be internationally competitive in cellular telephone technology it was going to have to achieve a higher level of circuit integration (i.e. pack more circuits on each chip) and offer advanced human interface and software features that competitors did not have. As well as improving product size and performance characteristics, higher levels of circuit integration means a lower labour input and thus lower costs because several discrete components can be replaced by one chip. In order to achieve the target levels of integration the right technologies are required, for example Computer Aided Design (CAD) for VLSI. Alvey was a happy coincidence because it was supporting R&D in the enabling technology areas, particularly VLSI and human interface, that Racal required.

In addition to the above cellular requirements, Racal Research personnel had interests in several projects which related to their core business but were less central. For example, a Racal senior manager was interested in remote terminals and 'electronic notebooks', and a research manager was interested in developing a route guidance and traffic information system for drivers. Discussions were held with five potential industrial collaborators, as well as with the Alvey Directorate. A composite large demonstrator project began to emerge under the title 'mobile IT terminals operating over cellular radio'. This included three sub areas: advanced cellular radio; mobile electronic office (MEO) to develop data networks and terminals operating over radio; and the vehicle route guidance system.

Following a feasibility study by Racal, Alvey requested that the project should incorporate R&D across all the Alvey enabling technology areas, including Software Engineering and Intelligent Knowledge-Based Systems (IKBS). Surrey,

Sussex and Cambridge Universities became involved. These were new collaborative partners for Racal, although the firm had a long previous history of collaboration with academic researchers.

Securing industrial partners was not without problems. One potential partner was already involved in a feasibility study for the route guidance element, and although its researchers were keen, its senior management decided not to pursue the collaboration. Another interested organization was not thought to be a suitable partner. One overseas company was very keen to collaborate in the project until a change of Managing Director forced it to drop out.

Collaboration with Hewlett–Packard was facilitated by personal contacts between it and Racal–Milgo. Racal and Hewlett–Packard employees put the proposal together for 'Locator' which was to research the problem of tracking moving vehicles for mobile communications. Racal–Milgo was keen on pursuing work on the X.400 communications standard which was at the time seen as a key enabling technology, and was an area in which Hewlett–Packard was already working.

In early 1985, whilst negotiations were continuing, the project was reviewed by Racal following very bad financial results for the company; the first year for thirty years that Racal had not made a profit. When the project was originally set up, Racal was able to take a long-term view, but pressure from the stock market and shareholders forced retrenchment. The decision was taken to concentrate on Racal's core business (cellular radio) in the project. The route guidance element of the project was terminated, as Racal could not see a way of recouping profits from it, and because of lack of support from other UK firms. However, the locator project survived in a modified form, concentrating on X.400 and X.500 standards and security aspects of mobile communications. For the remote terminals element of the project, Racal needed to collaborate with a personal computer manufacturer because this was not an area any Racal company was involved in. Racal approached a computer company to become involved and produce the required PC terminals. The collaboration came to an end when this company got into financial difficulties.

In general, Racal management were surprised at the reluctance of UK firms to take advantage of the 50 per cent government funding which they perceived to be an important incentive. These problems led to Racal seeking new collaborators through 1985, and a major restructuring of the project in 1986. The decision to restructure was taken at senior management level in Racal. MIS was reorganized into two major projects, and two minor sub-projects. The two major projects were:

1 the Advanced Cellular Voice/Data Demonstrator (ACVDD), to develop a hand portable cellular radio for voice and data transmission, involving two other companies and a university department;
2 the Locator project to develop secure electronic mail networks for mobile users, involving Racal Research, Racal–Milgo and Hewlett–Packard and a university department.

The minor projects were:

1   Traffic Information Collator (TIC) to develop a system to analyse traffic reports and prepare messages for broadcasting, consisting largely of research at Sussex University;
2   Fault Diagnosis IKBS (FDIKBS) for use in electricity supply networks, involving collaboration between two universities and a regional electricity board.

ACVDD was finally started at Easter 1986, and 'Locator' completed its definition study in January 1987. Project effort was concentrated on the two main sub-projects. The Fault Diagnosis IKBS sub-project had a budget of around £0.4 million and TIC only around £50,000.

Racal's only involvement with the Fault Diagnosis and TIC sub-projects was via their role as overall project managers. The negotiation of intellectual property rights (IPR) collaboration agreements, and the agreement between Racal and Alvey was time consuming, as was the case with many Alvey projects, since these were left to the individual partners in each project. However, there have been no subsequent IPR problems and none are envisaged in the future.

At the peak, Alvey work represented around one-fifth of the overall R&D programme within Racal Research Limited. However, Alvey work accounted for most of Racal Research R&D within the specific project research areas.

Relations with Hewlett–Packard were helped by the presence of an ex-Racal employee as one of the main Hewlett–Packard researchers. Collaboration with Hewlett–Packard in 'Locator' was successfully effected by both teams working on their own approach to the research tasks and then pooling knowledge and helping each other over barriers. The dual approach was appropriate because the work was concerned with standards, and so the two systems provided an opportunity to establish what they could interface. Hewlett Packard was able to get ahead of Racal at first because of pre-existing expertise. In the work on cryptography for 'Locator' both Racal and Hewlett–Packard needed to acquire expertise in the area, and there was some productive rivalry between the teams.

For Racal, management of the whole MIS Large Demonstrator project was not without difficulties. There was little linkage between the four disparate sub-projects. Progress within modules was largely independent and each element of the project needed to be managed differently. For example, the project manager met with the Sussex University team in a formal meeting every six months, but the academics essentially ran themselves, producing quarterly progress reports. The Fault Diagnosis sub-project met every six to eight weeks, and locator every month or so. Good interpersonal relations were argued to make a lot of difference to the running of the project.

Communication was considered to be all important. The project manager encouraged all R&D teams to circulate project reports as fast as possible in order to speed up the transfer of information within the project. Alvey had created an electronic mail system which was considered invaluable in assisting communication. Alvey electronic-mail was used to produce the original 'Locator' feasibility study between Hewlett–Packard, Racal Research and Racal–Milgo. Communication was argued by some to be better at times between Racal Research and Hewlett–Packard than with other Racal operating units that were not using electronic mail.

Difficulty recruiting researchers, particularly in the case of academic teams, was a major problem across the Alvey programme. This had some adverse effects on the MIS project.

Following the Alvey Directorate's intervention, a formal methods software development component (i.e. mathematics-based software specification) was included in the project. However, this was thought to be inappropriate for a demonstrator project. Some time was lost in attempting to use formal methods, which were in the end thought not to be viable for this application, being too research-oriented.

Shifts in corporate factors beyond the control of R&D managers, and changes in senior management over the course of the five-year project had a considerable impact on research direction. Racal suffered the loss of a key individual – a project 'champion' – on a number of occasions, which brought in new personnel with different interests and priorities.

The Alvey Directorate intended that the Large Demonstrator projects would have guaranteed access to all Alvey enabling technologies across the whole Alvey Programme. In practice there was no infrastructure to support this, and Alvey project participants protected their IPRs. The Demonstrator projects had to buy the technology in from the R&D projects on a commercial basis. In spite of the basic intention that MIS should take up research from other Alvey projects, there was no formal mechanism for technology transfer, and making links was difficult. Links developed on an *ad hoc* informal basis, but these were accidental and unsatisfactory. The second objective for the Alvey Large Demonstrator projects was to provide application-led goals for the R&D projects. This was found in the MIS project to be difficult.

However, Racal participated in other Alvey projects, particularly in VLSI and CAD (for chips) and through these, links were developed between MIS and Alvey VLSI R&D at Ferranti (later Plessey) Semiconductors. Racal Redac's participation in Alvey CAD R&D has linked the MIS demonstrator work to advanced CAD technology.

Racal considers that it has benefited from the collaboration, and will be exploiting some of the technology in its major product areas. However, as with all enabling technology R&D, the benefits accrue in terms of technology and knowledge diffusion into many products and systems over a long period.

The ACVDD sub-project greatly benefited from the European Community announcement of the Pan-European Digital System (PEDS). Digital data transmission over cellular devices allows lower power, cheaper equipment than analogue ones. The timing of the pan-European system was perfect; the Racal ACVDD team was set up for Alvey, then PEDS came along and the project was able to re-deploy quickly in order to meet this need and gain a strong presence in Europe. The MIS project gave Racal an edge they would not have had but for Alvey, and their presence has influenced the development of the PEDS system, and the development of European and world telecommunications standards.

A semiconductor collaboration with Ferranti enabled Racal to develop detailed specifications for the ACVDD chips. Racal values highly the benefits from its semiconductor collaborations. The Racal company Orbitel has benefited from the ACVDD R&D which has led to its being first on the market with a new type of telephone in 1990.

The 'Locator' sub-project developed secure multimedia electronic mail systems which will work on the pan-European network, and Racal and Hewlett–Packard contributed to the international standards. Racal would not have pursued R&D into X.400 if Alvey had not funded 'Locator', and indeed the Racal work may have been cut during the project because of financial pressure, if it had not been collaborative R&D. Collaboration thus has the advantage, to R&D departments, of locking firms in to R&D and assisting survival in volatile investment climates. Such locking-in would, of course, be seen as counter-productive from a business point of view if the R&D no longer fits firm strategy.

Racal Research considers that security technologies for X.400 systems are enabling technologies which will prove essential in the future, especially for EDI (electronic data interchange). Racal has gained expertise in encryption and influence on international standards development through participation in MIS.

Alvey set the project up and helped assemble the teams. In providing a collaborative framework, Alvey helped break down commercial barriers earlier than would have otherwise been possible. Alvey R&D in other projects involving Racal, especially the CAD work, enabled the Large Demonstrator project to move further and faster than would otherwise have been the case. Racal has also learned a lot about collaborative R&D. It is participating in two new DTI funded R&D projects, and is involved in ESPRIT semiconductor research.

The following reasons are suggested by Racal managers for the success of the collaboration:

(a) good interpersonal relationships have been very important in pulling a diverse project together;
(b) good communications between the teams has been essential;
(c) project management has needed to be very good;
(d) although very time consuming, sorting out the ownership of IPRs early on proved very valuable;
(e) timing is important: the Alvey project arose at a time when Racal wanted to get into the new area.

The above story only looks at the situation of the project from Racal's perspective. It describes the many problems involved, such as coordinating different organizations with differing and changing objectives. One of the major problems in assessing the impact of any collaboration is the varying perspectives of participants. It does not, for example, pose the question of whether the project attained its public policy objectives. Racal's eventual focus in the project was very different from the initial objectives, and could conceivably be argued to be nearer-market product rather than research oriented. While the company individually benefited from the project, the question must remain of whether success can be claimed from the perspective of the original aim of the project.

CONCLUSIONS

ESPRIT, Eureka and Alvey have all had generally positive impacts for participating firms. They have opened new communications paths and have allowed firms to access new capabilities. Evaluations of these programmes have also revealed many of their shortcomings. Whilst it is a key intent of these programmes to encourage participation by smaller firms, such firms find it difficult to participate. Indeed, it has to be asked whether collaborative R&D is the most appropriate form of public policy innovation support for smaller firms (Garnsey and Moore 1992). The degree of additionality of programmes such as Eureka is questionable. Many of the projects built on existing inter-company links and would have been undertaken without public support. In many of the Alvey projects, companies sought public funds for projects central to their future which otherwise were unlikely to be undertaken.

Some of the complications of these schemes have been highlighted in the case study, such as the balancing of divergent requirements from a public policy and corporate perspective. In the case study, the demands from the Alvey Directorate and senior Racal managers were occasionally contradictory. Circumstances change in partners and within the firm. In the case study, for example, changes in senior management affected attitudes to the Alvey projects. The objectives and structures of the projects were altered considerably during the course of the programme, and these complexities are compounded by issues such as ownership of IPRs.

Despite all these problems for both public policy-makers and companies, the in-depth evaluations of collaborative programmes described here are all of the opinion that, in balance, they have made a positive impact. Such a verdict is made, of course, in the context of economies and technologies which are severely disadvantaged compared to the world leaders.

Chapter 8

Corporate technology strategy

Technology is increasingly accepted to be a critical issue for corporate strategy (Dodgson 1989; Loveridge and Pitt 1990). Such strategies are concerned with the accumulation and exploitation of technology. Collaboration can play an important role in both these areas. This chapter argues that an aim of technology strategy is to develop useful capabilities around the core activities of firms which are affected by changing and uncertain technologies, and which can assist in technological diversification. The development of these capabilities requires firms to learn, and this includes learning from mistakes. Two case studies are provided of the ways companies may fail to achieve the original objectives of collaboration, but in doing so can usefully supplement their capabilities which can then be re-directed.

TECHNOLOGY STRATEGY AND COLLABORATION

The process of technological accumulation has traditionally been seen as an internal process. Mytelka (1991) argues that most contemporary theorizing on multinational companies, based on a transaction cost approach, emphasizes their preference for internalized hierarchies rather than market solutions. There are, however, countervailing tendencies. For Chesnais (1988),

> The complexity of scientific and technological inputs, the uncertainty of economic conditions and the risks associated with uncertain technological trajectories, appear to have reduced the advantages of vertical and horizontal integration and made 'hierarchies' a less efficient way of responding to market imperfections.

> (Chesnais 1988: 84)

To maintain their long-term competitiveness companies need continually to improve the technologies which provide the core of their activities, and to adjust to the new and emerging technologies which can affect their core areas and perhaps provide potential future central activities through diversification. Building these core competences requires technology strategies that involve complex decisions concerning which technologies are appropriate and necessary for the firm's long-term profitability and growth and how these are to be developed, accessed and diffused.

External inputs into technology strategy formulation and implementation are highly important. Collaboration can assist the decision-making process about the potentials of novel technologies. Assessment of current and future technological strengths and weaknesses within the firm can be assisted vicariously through observation of other's capabilities in collaboration and technologies new to the firm can be developed and accessed through collaboration. As the success of the Japanese firm NEC shows, high numbers of collaborations, if managed effectively with a strategic perspective, can help access critical new technologies quickly and cheaply, and can facilitate business diversification (Prahalad and Hamel 1990).

Technological collaboration is in major part a result of the uncertainties engendered by contemporary technological change.

> It becomes essential to relate to the behaviour of firms in complementary horizontal and vertical activities, as the new technologies provide wider opportunities for those firms to affect competitiveness. To overcome the problems of complexity, high cost and high risk, activities previously proprietorial to individual firms such as R&D and manufacturing may become shared between a number of firms. The necessary sacrifice of autonomy in the generation and diffusion of technology involves a strategy of sharing control in order to retain it. Without participation in multilateral technological arrangements, even the most advanced companies may lose their leadership positions.
>
> (Dodgson 1989: 6)

Granstrand and Sjolander (1990) conceptualize technology strategy as a way of adding to and exploiting the technology base or capability of the firm. Joint ventures and inter-firm R&D cooperation are seen as elements in both sourcing and exploitation strategies, and contract R&D is also seen as a means of technology purchasing and selling. Technology strategies are increasingly complex, they argue, because firms are multi-technology. The range of technologies many large companies are involved in are shown diagrammatically in Table 8.1 (van Tulder and Junne 1988).

Granstrand (1991) proposes that the focus of strategic management tasks of the multi-technology corporation are extensive, and many have relevance for collaboration, including:

(a) External technology sourcing (acquisitions, joint ventures, licensing, contracting, technology scanning);
(b) technological economies of scope (combination, confluence, fusion of technologies);
(c) avoidance of the 'not-invented-here' syndrome;
(d) technology-based product diversification.

Harrigan (1986) suggests a number of strategic uses of joint ventures, including increasing or decreasing the range of a firm's activities, and stabilizing a firm's

Table 8.1   Technology diversity in large European firms

Company	Chips	Computers	Robots	Telecoms	Software	New materials
Philips	*	*	*	*	*	*
Siemens	*	*	*	*	*	*
Daimler/AEG	*	*		*	*	*
CGE		*	*	*	*	*
GEC	*	*	*	*	*	*
Thomson	*	*	*	*	*	*
Ericsson	*	*		*	*	*
ICL/STC	*	*	*	*	*	*
Matra	*		*	*	*	*
IRI/STET	*	*	*	*	*	*

Source: van Tulder and Junne (1988)

Note: CGE and Thomson are now merged

existing activities. They can achieve a variety of diverse strategy objectives, such as:

(a) creating synergies between firms;
(b) enhancing innovation in a specialized and flexible manner;
(c) allowing firms to diversify into attractive but unfamiliar business areas, and diversifying from unfavourable businesses into more promising ones;
(d) expanding business internationally;
(e) divesting businesses which no longer fit corporate objectives.

Internationalization is a key theme of much of the corporate strategy literature (Casson 1991; Doz 1986). Technological collaboration can play an important role in the pressures towards internationalization facing many companies. For example, Imai (1990) relates how Toshiba is developing a global network through many strategic alliances.

> To the management of companies confronted with the need for fast internationalization and related diversification, strategic alliances and partnerships provide an interesting option to access the world markets and/or combine with adjacent technologies. Companies that have been primarily domestic (e.g. AT&T in the US, GEC and Plessey in the UK) may see in partnerships a short cut to world markets. Companies affected by converging technologies (e.g. Olivetti and AT&T, NEC and Bull) may also see in partnerships a short cut to full coverage of related markets. Partnerships provide low cost fast access to new markets, by 'borrowing' the already-in-place infrastructure of a partner.
>
> (Doz, Prahalad and Hamel 1989: 121)

An important strategic consideration in technological collaboration is the question of linkages between large and small firms. Collaboration may provide a means by which the 'dynamic complementarity' (Rothwell 1983) between the resource advantages of large firms and the behavioural advantages of small firms can be achieved. In Chapter 5 an example of strategic linking between large and small firms was described. Small companies such as Celltech have grown on the basis of their technological collaborations. And large firms can similarly benefit from inputs from small firms (Arora and Gambadella 1990). However, as will be examined in greater depth in Chapter 11, large firms tend not to be particularly expert in their strategic linkages with small firms.

One option for a large firm wishing to access the innovative capabilities of a small firm is acquisition. However, as Mowery (1988) argues, collaboration is often used for corporate diversification because of the high failure rate of mergers. This applies also to large firms' acquisition of small firms. Roberts and Garnsey (1992) show how, after a period following acquisition, smaller firms begin to lose the entrepreneurial edge which attracted large firm buyers. Collaboration provides an alternative to acquisition, and in the case of small firms allows them to continue to develop their specialist technological capabilities which are the main attraction to large firms.

Two case studies are now presented: the first describes the formation of a joint venture, with one partner having the aim of diversification, the other aiming to exploit existing know-how. The second case study describes a collaboration entered into to increase the linkages between two firms in the same group of companies, which proved technologically successful, but unsuccessful in a business sense.

CASE STUDY: BT&D TECHNOLOGIES LIMITED

BT&D Technologies Ltd (BT&D) is a joint venture formed in 1986 between British Telecom (BT) and Du Pont. BT&D researches, develops and manufactures a range of optoelectronic components for the global telephony and data communications markets. The company employs 300 people in the UK and another fifty in the US. Turnover increased from £700,000 in 1988 to £5 million in 1989, and roughly £10 million in 1990.

In the 1980s, Du Pont was concerned to expand beyond its traditional and relatively mature core businesses of chemicals, petroleum and materials. It identified electronics as a critical market and set up an electronics division which dealt principally in electronics related materials. The company developed a strategy to become a major international electronics concern.

Du Pont was interested in developing electronic devices, building on and complementing its materials and manufacturing competences. The company had a history of success in diversification and partnership, and a joint-venture was felt to be the most appropriate method of acquiring product knowledge.

The impetus for the push into optoelectronics was commercial. Du Pont believed that the market for optoelectronic components would grow massively in the latter part of the 1980s and into the early 1990s with the anticipated development of fibre-to-the-home for telecommunications.

The reason for BT's interest was similarly commercial, but it was also driven by an untapped technological capability within the company. BT had moved from public ownership into the private sector in 1984 and was concerned to find a commercial outlet for the vast stock of leading-edge technical knowledge it possessed in its British Telecom Research Laboratory (BTRL).

BTRL was a pre-eminent technical research organization, but it was inexperienced in commercial activities. A joint venture was believed to be the most appropriate solution. A successful joint venture which generated revenue from some of BTRL's optoelectronic activities would justify BTRL's role and continued existence within BT. It would also provide BT with a leading-edge component manufacturing capability, which was in line with BT's strategy at that time.

Du Pont began looking for a possible partner in 1984. Its first choice was a large USA telephone company. Du Pont also looked around Europe, but the decision to work with BT arose from a chance meeting of BT and Du Pont chairmen at an industry conference in 1985. It was quickly established that there was a complementarity of both capability and interest between the two companies. Du Pont was a leader in electronic materials research, engineering, manufacturing processes and global marketing, while BT had a technologically world-class fibre optic and optoelectronic capability. Both companies wished to develop an optoelectronic business. The fit between the two was further enhanced by the fact that they were not competitors, and had no intention of becoming so. The structure of the collaboration has evolved over the years. The way that it has been set up and supported shows the high levels of commitment to it by the parent companies.

The venture was capitalized originally on a 50:50 basis by the parent companies. Actual funding was determined on an annual basis with BT&D directors taking responsibility for justifying the next year's budget. In total, by 1990 some $50 million had been spent on plant and facilities, with a further $50 million invested in R&D. One-quarter of total funding, however, has been spent with the parent corporations: BTRL for device specific and research elements; Du Pont for production machinery, and some packaging and materials. With this arrangement, BT&D assumed the role of contractor and, as such, controlled the development process.

In order to strengthen the new venture as much as possible, both BTRL and Du Pont waived the rights to commercialize parallel proprietary R&D. All optoelectronic inventions are marketed through BT&D, which pays royalties to the originator on a product-by-product basis. BT&D even has the option to license such developments to third parties.

BT&D initially chose a high risk strategy of developing leading-edge products. Du Pont had provided BT&D with market intelligence predicting a generic, high volume market in optoelectronic components built on fibre-to-the-home. Based on this, BT&D drew on the breadth and depth of BTRL's technical expertise to design and develop a system with sufficient performance to meet a wide range of technological requirements. The result was a high technology system based on indium gallium arsenide technology which is claimed to have 'scared the hell out of everybody in the business'.

However, while the new receiver out-performed everything on the market, it was difficult to make and it was very expensive. Ultimately, only 5 per cent of BT&D's business needed such high levels of technological sophistication. Furthermore, the expected market failed to emerge, and indeed is now considered unlikely to do so within the next ten years.

These potentially disastrous miscalculations actually had some positive impacts on the company's development. The fledgling company had the opportunity to display its preeminent technical expertise, and it established considerable credibility with major large customer firms in the process. This, plus BT&D's association with Du Pont and BT, garnered a reputation within the industry out of all proportion to the company's size and age. This proved important, as the company was able to use the technological capability it had developed to address new markets. In particular, it has enabled BT&D to adopt the marketing strategy of a focussed solutions supplier. It targets blue chip large firms, and at present five major customers account for 80 per cent of BT&D's business. The company has also benefited from an unforeseen growth in datacoms markets.

BT&D's original business plan considerably underestimated the size and timetable of the parent companies' investments. Furthermore, since the joint venture was originally planned, BT and Du Pont have undergone changes in operating circumstances. Immediately after privatization, BT had ready access to finance and was committed to developing a manufacturing capability. By 1989, it was operating under severe financial constraints and had revised its corporate strategy to concentrate on being a service provider. The resurgence of the chemical and materials industries in the past few years could also potentially affect Du Pont's diversification strategy.

BT and Du Pont have, nevertheless, continued to be supportive of BT&D. BT's interest, in the wake of a strategy review, was largely reputational and financial. Du Pont, however, remains intent on becoming a major electronics force and BT&D believes it has already provided them with invaluable expertise at product level. In 1989, when the company requested additional financing, Du Pont took the opportunity of increasing its share of the venture to 60 per cent.

To date, most of BT&D's revenues have accrued from technologies and knowledge transferred from the parents. The long-term intention is for BT&D to internalize much of this expertise and build its own basic R&D core skills. This build-up is intended to be market driven, and BT&D will continue to commercialize parent company technologies.

The interdependence with BT and Du Pont is seen as a desirable element of BT&D's future business. BTRL is seen as a valuable source of new products which can be developed initially for niche markets, and subsequently marketed more broadly. The company's advanced products department, for example, has taken ideas directly from BTRL to produce small volumes of leading-edge products for known customers. Within a two to three year timeframe, the more successful of these have evolved into volume product lines. This process is planned to continue.

Executive management of BT&D lies with a Board of Directors consisting of two BT and two Du Pont senior employees, and BT&D's Company Secretary. The Board is chaired by BT&D's Managing Director, who does so in a non-executive capacity. BT&D's first Managing Director was a Du Pont secondee. In early 1990 he was succeeded by a manager with expertise in running start-up companies, and with lengthy experience in the USA optoelectronics industry. A major part of the new Managing Director's brief was to develop a corporate identity and managerial systems separate from the parents'.

Within BT&D it is believed that building a working relationship with BT has taken some time. The parent organization itself has had to go through a considerable learning process about collaboration and commercially orientated start-up ventures. In contrast, Du Pont had built many start-ups and with this experience had consequently been better able to cope with BT&D's initially faltering performance and short-term fluctuations in fortune.

BT&D's Managing Director continues to report to BT's Director of Science and Technology, responsible for BTRL and BT's procurement, who in turn reports to the BT chairman. In Du Pont's case there is a more direct link through the Director of the electronics division. BT&D found itself reporting to a service function on BT's side, and a business function on Du Pont's.

The formal mechanisms and interfaces between BT&D and its parents were allowed to evolve over time. During the initial technology transfer stage, BT&D employees, who were mainly seconded from the parents, had dual commitments: to the parent and to the venture. The main formal links were through project managers back into both parents' labs and development areas.

Given the different cultures and experiences of the three organizations, tensions inevitably arose in the early years. BTRL secondees had found difficulty in adjusting to the new commercial regime and to market-driven research. Also, BTRL employees had been rather suspicious of Du Pont and Du Pont secondees because of their perceived lack of electronics know-how. The novelty of working within a commercial environment, and in collaborating in complementary areas rather than with peers, had meant that BTRL secondees had found the joint venture more difficult than their Du Pont counterparts. For their part, the Du Pont secondees had struggled to move beyond the 'Du Pont way' and the big-company ethos. In the early years there was duplication in certain functions, such as personnel and administration. Although the personnel professional from BT&D was to oversee the parents' functional representatives, there were differences between 'American' and 'British' employees.

In an attempt to resolve these confusions about responsibilities and loyalties, BT&D's new Managing Director established and enforced clearer BT&D objectives, reporting structures and responsibilities. In 1990, the links back into the parent companies were more formalized and the company began to report to BT's product strategy committee. The majority of secondees have been repatriated. An area that

BT&D had done a considerable amount of work on was the development of a clear set of design rules. The biggest problem here had been to convince researchers of the need for a rigorous design procedure in order to move research quickly and smoothly into the prototyping and manufacturing phases.

BT&D is now in a position to begin to take commercial advantage of its own recently developed competences, built on the basis of those of its parents. Its latest business plans, according to managers within BT&D, are more realistic than previous ones, and begin to chart how the company can capitalize on its technological strengths in a new and growing market. It indicates that BT&D needs continuing financial support from the parents. However, it is believed that if BT&D is to continue to grow rapidly it is going to have to become increasingly independent. It will have to learn how to operate as a separate company. BT&D's future depends to a great extent on the way it can develop its own identity and develop its individual practices and procedures conducive to success in its own particular market.

BT&D's original business plan was misguided in a number of respects. BT and Du Pont had little experience of the electronics industry, and an unrealistic growth pattern was expected for the joint venture and for the market in which it was to operate. As forecasts were not met, morale in BT&D suffered, and there was disquiet amongst the Joint Venture board. BT&D's internal confidence suffered even though the company's growth performance was well above average by other industry standards.

A second problem was related to the differing cultures of the organizations involved. Initially there was a confusion of identities at BT&D and a lack of clarity in reporting structures and systems. The cultural mindset of BT&D was that of a large company: departments constantly pushed for expansion of facilities and personnel in order to meet notional growth rates. As a result BT&D had grown faster and spent more than was appropriate to its circumstances.

BT&D's own understanding of how to manage a collaborative venture and control the relationship with its parents had unfolded only slowly. With hindsight, BT&D managers feel that collaborations need to establish their sovereignty much earlier through visibly independent structures, systems and culture. It needed to be established early on that employees' loyalties had to be with BT&D first and the parent organizations second. It is felt that this independence would have been facilitated more quickly had an independent, industry specialist been appointed as Managing Director earlier.

The following, in no particular order, are suggested in BT&D as being reasons for the success of the joint venture:

(a) BT and Du Pont had clear objectives. Du Pont wanted to diversify; BT wanted to commercialize BTRL research. Both parents were determined to make the venture a success.

(b) The partners had complementary interests and capabilities. Both were, at the beginning, cash-rich, and BT&D received high levels of investment.

(c) The joint venture benefited from its high-level origins which engendered high trust from the outset. Indeed, the company's site was selected, and refurbishment begun, on the basis of the chairmen's handshake and prior to the agreement being finalized.

(d) Du Pont had considerable experience of forming joint ventures. At the outset, Du Pont's International Department had taken responsibility for coordinating much of the work whether that was refurbishment, plant procurement, personnel or safety. Du Pont had established systems that BT&D were able to utilize to get the joint venture underway quickly and smoothly.

(e) Initially, in order to commission the facility and give the company a strong base from which to work, BT&D chose to recruit a select number of highly qualified people, rather than going for large numbers in a rapid search for high output levels.

(f) Another characteristic of success was high levels of communication. This was essential as the joint venture was working across two corporate and national cultures. BT&D was committed to keeping lines of communication open. Again, Du Pont's earlier venturing experience and existing international mechanisms and systems assisted this process.

(g) Timing was fortuitous. The time was right for both partners to form the venture. Indeed, their perspectives on forming such a venture might prove very different today. Also BT&D were fortunate to find an alternative market emerging at the time when its original market proved inappropriate. The capability of responding to this change depended, however, on the company's strength of technological and financial resources.

Another example of technological collaboration resulting from corporate strategy considerations is provided in the following case study. In this example, which was technologically successful, but unsuccessful in a business sense, some of the problems of collaboration are highlighted.

CASE STUDY: QUANTEL/SSL

Quantel develops, manufactures and markets image processing equipment, primarily for the television industry. It enjoys a reputation for producing highly advanced innovative products, several of which have, in a short period, become standard within the industry. It currently employs over 600, and is part of Carlton Communications plc.

Solid State Logic (SSL), a sister company of Quantel (both were in the UEI group of companies acquired by Carlton Communications in 1989) produces equipment for the professional sound market. In early 1987 Quantel and SSL began a collaborative project aiming to merge Quantel's digital image processing know-how with SSL's sound processing expertise. The intention was to produce a joint product which processed both sound and images. Rather than being formulated in line with SSL's existing products of large consuls with hundreds of dials, the new product was to follow Quantel's product configuration of a screen and drawing tablet.

Quantel is the technological and market world leader in its field of image processing. Products such as 'Paintbox' and 'Harry' have created new technical possibilities for the television industry and are now almost considered essential by television companies. The company is constantly updating its products and in 1987 it was concerned to add a digital sound editing function to its image processing product, Harry.

Within Quantel there is little enthusiasm for technological collaboration. The company's management is highly aware of all the potential problems involved in collaborating in R&D. It is felt that as the company is the best in the field, there are likely to be few potential partners with comparable levels of expertise. Nor is it believed that many other companies can match the high levels of motivation that Quantel enjoys amongst its fifty R&D staff. The company does collaborate with a number of Japanese companies, but these usually involve manufacturing to Quantel specifications. It also has a project on the Eureka programme. Generally, however, Quantel tends not to collaborate and a number of circumstances led it to do so on this occasion.

One of these was that SSL was a company in the same group. With its expertise in digital processing, senior managers within Quantel became aware that SSL's internal efforts to develop a product using digital processing had some serious shortcomings. It was felt that a collaborative project directed towards creating a sound 'add-on' to Harry would redirect SSL's efforts. Furthermore, it would enable SSL to sell products from a platform of existing Quantel customers, and therefore learn about the market.

Negotiations on the collaboration were undertaken by Quantel's chairman. The collaborative project was managed by an ex-Quantel employee, working at SSL, who was familiar with Harry. Quantel put a development team of three to work on the joint project, although this was not their only project. SSL had a similar complement of engineers working on the project. The proposed name for the product was HarrySound.

None of SSL's engineers working on the project were amongst those half dozen R&D engineers who were working on the digital processing of signals prior to the collaboration. It was Quantel's concern over the direction in which this research was moving that was one of the major reasons for the collaboration. When, these engineers later left SSL, Quantel's concern seemed justified.

The use of a screen and tablet to edit sound was novel to SSL, and indeed there was some scepticism within the company as to whether it was a viable alternative to existing control panels. This was, however, soon overcome. One of the greatest technical challenges to using digital editing on tablets lies in the ergonomics and associated software. In this field, SSL was almost entirely reliant on Quantel's expertise. Quantel provided SSL with the kind of control software it used in its systems. This amounted to a considerable and expensively accumulated resource.

After twelve months or so, a product emerged from the collaboration. Quantel was pleased with the technical progress on the project, and believed that the product had met perhaps 80 per cent of initial specifications. It considered that the remaining 20 per cent of development work was necessary in order to ensure that the sound add-on to Harry gained widespread market appeal. SSL's management, however, decided that, given the company's position at the time, the extra work was not to be undertaken, and it began to market the product. Named ScreenSound, this product is currently being sold on the digital audio market.

Quantel had not attained what it wanted from the collaboration namely, a sound add-on to Harry. It continued to work with SSL for a period of around six months, but then decided to pull out and to develop its own system. Quantel began developing its own system in November 1989, and its resulting product, Harry-Track, was first launched in April 1990.

The way in which Quantel approached the collaboration was dictated by its relationship with SSL as a sister company in the same group. This is reflected in its preparedness to contribute such a major proprietary asset as its software. Considerations of the equitable trading of know-how and of possession of intellectual property rights were not in the forefront.

The operational management of the collaboration reflected the inter-personal knowledge of the researchers in both companies. There were few formal review procedures involved. Both teams worked together well. Communications were assisted by the two partners being computer linked. However, the manager responsible for the project on Quantel's side found the delays in communicating with another company to be a problem. Until the period when the product reached what Quantel believed to be 80 per cent of capacity, the project was successfully managed in that technical progress was considered satisfactory, and in terms of its engineering, the system worked well.

Obviously the major problem the collaboration faced from Quantel's perspective was the differing objectives of the two partners. Quantel wanted a piece of equipment to add to its range. SSL wanted its own product. Quantel's own objectives were two-fold, relating to its own needs, and to its role as a part of a group. As far as the strategic objectives of the collaboration were concerned, therefore, there was no simple, unified intention.

At an operational level, one of the major problems experienced by Quantel's project manager was the communications process, despite the fact that the communications paths between the two companies were clear and based on good inter-personal linkages. It was found that the length of time communicating Quantel's objectives, gaining agreement over them, and getting them acted upon was a problem to a company used to quick responses.

Assessing how successful the collaboration has been to Quantel depends on how its objectives are viewed. From its own parochial perspective its requirements had not been met. Quantel devoted one to two man-years of development time to the project, and gave away some of its valuable software in the process with no tangible result. From its perspective as a member of a wider group it perceived that it had helped SSL redirect its R&D efforts, and had enabled the company to develop its skills in digital audio processing.

Within the company it is believed that Quantel lost perhaps eighteen months in its efforts to get a sound function on Harry through its collaboration with SSL. However, a number of learning processes occurred during the collaboration which proved important to the company. These are reflected in the speed with which Quantel developed HarryTrack, and the way this product has met almost perfectly what was demanded of it by the market.

HarryTrack took six months to develop within Quantel; from the formulation of the product specifications to its market launch. Particularly useful in this development process was the knowledge of the needs of customers for the product. Much of this knowledge was obtained through feedback from customers during the collaboration. One of the lessons learnt during this process was that initial specifications for the product were technologically over-sophisticated. HarryTrack's final configuration reflected this lesson.

Quantel had little experience of sound processing technology prior to the collaboration. As a senior engineer in Quantel put it, 'through our collaboration with SSL we learnt what we needed to know in the area and, importantly, we learnt what we did not need to know'. When entering a new technological area, knowing what sorts of information are appropriate gives a number of advantages. It prevents time and effort being wasted in the search for information which turns out to be of little use to the project.

Although Quantel's evaluation of the success of the collaboration shows limited results, the company benefited in a number of indirect ways. These are perhaps most clearly shown in the way that the market has welcomed the product, Harry-Track, which was technologically novel to Quantel and something of a departure for it. It managed to produce a product which fulfilled the demands of the market, and it got it right first time.

CONCLUSIONS

In Chapter 2 the question was posed: is technological collaboration a strategic issue? The answer provided by this chapter is an emphatic yes. While not discounting the relevance of using collaboration to access a piece of knowledge or technology for immediate, tactical reasons, collaboration can also assist the development of core competencies and technological diversification; both central strategic issues. Examples of NEC and Toshiba were cited as firms that had used collaboration to build their core competencies and diversify. In one of the case studies Du Pont was seen to be using joint ventures to diversify from its traditional markets. Collaboration provides an opportunity for firms to learn about new opportunities and ways of doing things.

Both the case studies reveal the way the outcomes of collaboration may be different from the ones initially expected. BT&D was directed towards a market that failed to develop as envisaged, and its first product was too 'high tech'. Quantel failed to develop the product it wanted through its collaboration. However, through the process of collaboration both companies had learned and developed *new capabilities*. These outcomes were successfully put to use in new markets and products. It is through the development of new capabilities that the strategic implications of collaboration become apparent.

Chapter 9

Internationalization and technological collaboration

There is considerable debate on the extent to which the technological activities of firms are international. For some, we are already operating in global economies where technological development is a worldwide concern. Others argue that the domestic technological activities of firms predominate. This chapter argues that although currently, international technological activities are limited they may be growing in importance. At present international technological collaboration occurs primarily within the 'triad' of Europe, the USA and Japan, and are limited in scope. Firms are increasing their international R&D for a number of strategic reasons, important amongst which are increased opportunities for learning. A case study of an international collaboration outside of the triad is presented, and apart from accentuating many of the lessons from the other case studies on issues such as the need for good communications and trust, it also extends lessons about learning.

INTERNATIONALIZATION OF TECHNOLOGY

A number of reasons can be suggested why firms may choose to internationalize their R&D. Their markets and manufacturing plants may be international, and R&D may be set up to support them. Foreign nations may possess unique R&D resources, or their governments may mandate R&D expenditure as a condition of foreign investment or procurement. A major feature of this internationalization is argued to be collaboration. One of the most enthusiastic proponents of the increasing globalization of firms (Ohmae 1990), argues that as the world is seeing converging consumer tastes, rapidly spreading technology, escalating fixed costs and growing protectionism 'Properly managed alliances are amongst the best mechanisms that companies have found to bring strategy to bear in global markets' (Ohmae 1990: 136).

Others also argue that a primary feature of the global strategies of firms lies in their collaboration and networking activities (Porter 1990). Devlin and Bleackley (1988) contend that 'Probably the greatest stimulus to alliance formation has been the emergence of global competitors and those corporations wishing to become global' (Devlin and Bleackley 1988: 19).

As described in Chapter 3, there is an argument that technology and internationalization are combining in 'techno-globalization'. Sub-contracting, licensing, joint ventures, R&D cooperation, and inter-firm agreements have been described as representing 'different forms of networking among firms – the most ubiquitous mode of techno-globalization' (Gibbons 1990). Central to the techno-globalization analysis is the view that:

> the internationalisation of science and technology has gone hand in hand with an increase in transnational networks and strategic alliances between enterprises as a means to competitive advantage in global markets, increasingly through the joint development of and access to technology. It raises major questions about the role and grip of government policies, and their relationship to the strategies of such enterprises.
>
> (Soete 1991: 76)

Others, however, are much more sceptical about the extent and form of globalization. Hu (1992), for example, suggests that stateless operations do not necessarily mean stateless corporations. He argues that corporations have a geographical centre of activity where ownership and control reside, profits are remitted, top managers selected, strategic decisions about innovation made, and where relationships with individual governments and tax authorities predominate. The exceptions to this he explains are binational companies, such as Shell and Unilever (British/Dutch) and ABB (Swiss/Swedish), and firms from small nations, such as Nestlé, Philips and Ericsson. Apart from these 'there are no multinational, transnational or global enterprises, only national firms with international operations' (Hu 1992: 6).

So there are conflicting views. But what empirical evidence is there of the extent of internationalization of R&D and the forms that it takes? Some of the strongest evidence suggests that firms' technological development remains an essentially domestic activity (Porter 1990; Patel and Pavitt 1991b). The assumption of the increased globalization of technological activities is severely questioned by Patel and Pavitt, who argue that technological development essentially remains within firms' home countries. In their study of the technological activities of 686 of the world's largest manufacturing companies they found that more than 80 per cent of these activities are domestic.

Furthermore, foreign participation in publically-promoted collaborative schemes is low (USA firms find it difficult to participate in European Community schemes, and vice versa, and access into Japanese programmes is notoriously difficult for foreign firms). A classic case of the dilemma of inclusion/exclusion is provided by IBM's attempts to join the EC's ESPRIT Programme. IBM is the largest actor on the supply side of several EC member states, it undertakes substantial amounts of R&D within the EC, and it is not clear whether it has a negative effect on the EC balance of payments. If IBM were to be included, then the EC would be supporting a US firm's research, and allowing it access to European research results. Of more than a dozen IBM proposals to ESPRIT, only one was accepted (presumably to prevent accusations of discrimination) (Arnold and Guy 1986). This sort of problem also arose when ICL, the British computer company, was 80 per cent acquired by Fujitsu. Its position in a number of EC collaborative schemes caused considerable argument.

On the other hand, the evidence supporting a strong techno-globalism role of collaboration is very shaky (Dodgson 1992b) – something which is admitted within the Technology Economy Programme. None the less, the world is changing very rapidly, and some firms are increasingly international in outlook (Howells 1990; Granstrand *et al.* 1990). Soete (1991), for example, describes the way that between 1987 and 1990 Japanese electronics firms set up twenty-one new research centres in the USA, six in Europe and six in Asia.

The 'international' nature of collaboration is questionable in respect to the numbers of actually participating countries. As Figure 9.1 shows, the vast majority (over 85 per cent) of agreements take place amongst the triad of the USA, Japan and Western Europe. The non-triad share of agreements is mainly covered by companies in South-East Asia working with triad partners. The figure, which is based on three databases, shows a general picture of the highest levels of collaboration within the USA and between the USA and Europe. Alliances within Japan and between Europe and Japan are generally at a comparatively low level. There is believed to be a problem in the under-reported number of intra-Japanese collaborations in the MERIT/CATI and FOR databases.

The question arises, for an increased internationalization thesis, not only of how collaboration may exclude non-triad nations, but also the focusses of collaboration. As Figure 9.2 shows, collaboration in development is much more common within rather than between the triad of global trading blocks (an exception is EC–USA agreements). Production and marketing agreements are much more commonly the focus of collaborations between the triad.

Van Tulder and Junne (1988) in a study of 193 alliances found that European firms use their alliances with US or Japanese firms as a substitute for cooperation with other European firms. They found that the bulk of intra-European cooperations are in R&D, while the extra-European cooperations are closer to commercialization. This is so in the auto-industry, for example, where international production sharing agreements are common. Lamming (1992) gives some examples:

> The international networking of the auto industry has led to extensive opportunities for assemblers and component firms to amalgamate production in partnerships and subsidiaries in order to gain economies of scale. For example: the joint venture Saab Automobile (50:50 General Motors and Saab) will receive Vauxhall engines from the new plant in ... England in 1993; GM plants in Canada, Australia, and South Africa receive transaxles from Isuzu and Suzuki plants in Japan (General Motor's partners).
>
> (Lamming 1992: 111)

Although evidence from existing international databases shows limited alliances formed outside the triad, a number of countries are seeing increased collaborative activity. For example, strategic alliances are believed to be of growing importance in assisting the internationalization and competitiveness of Australian firms (Scott-

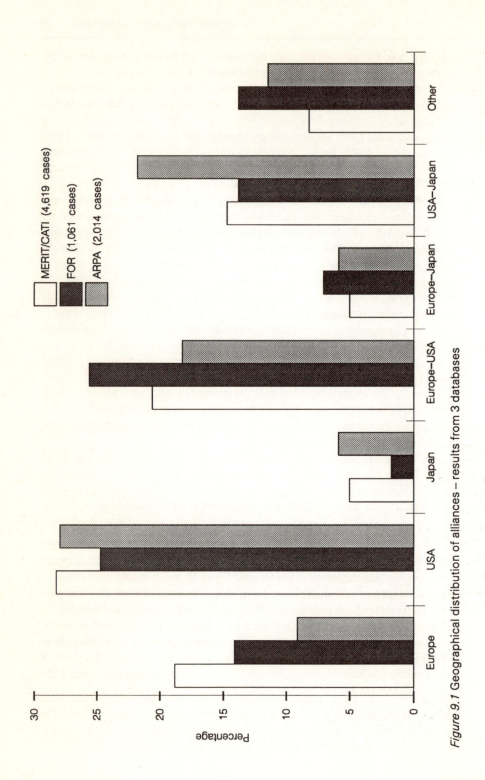

Figure 9.1 Geographical distribution of alliances – results from 3 databases

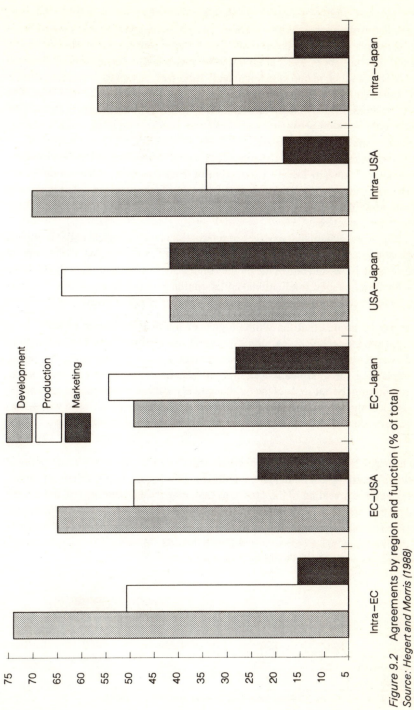

Figure 9.2 Agreements by region and function (% of total)
Source: Hegert and Morris (1988)

Kemmis *et al.* 1990). In a survey of strategic alliances it was found that for most Australian firms such activities were in their infancy, and that many small firms had particular difficulties collaborating with large overseas firms. However, it is argued that 'alliances were often valuable mechanisms for achieving strategic objectives, and in many situations had important advantages compared with alternative (market or internal) approaches' (Scott-Kemmis *et al.* 1990: viii). The authors recommended that government policy makers should prioritize the promotion of collaboration between Australian firms and firms in other small countries or small firms in large countries.

There is a rapid growth of collaborations within some of the old communist bloc countries, as shown in Table 9.1. For some of these countries collaboration is seen as an important tool in economic development. Technologically disadvantaged nations and firms may perceive international collaboration as a means of catching up with best practice.

The general picture that emerges of international technological collaboration is one that is at present limited in scale and scope, but this is changing. Collaboration is a strategic issue, and one of the themes of this book is that learning is a key issue of inter-firm linkages. International collaboration may increasingly reflect a strategy by firms to extend their learning. As De Meyer (1992) argues:

> It can be argued that the ultimate goal of internationalisation of R&D is to improve the technical learning process of the company's technical function. Faster learning of more relevant information is, in our opinion, the key to explaining the internationalisation of R&D. Learning about customer needs, monitoring the hot spots of the field to quickly learn about recent developments, and having access to resources (engineers, scientists) which can process this information quickly is the objective of the internationalisation process.
>
> (De Meyer 1992: 13)

An example follows of the ways in which collaboration can improve international partners' learning technological capabilities. This partnership is relatively small-scale, yet it is revealing of the problems that arise in international collaborations, and particularly of the importance of good communications flow and building trust between partners. It is not a collaboration within the triad (the focus of the majority of the literature, see Contractor and Lorange 1988; Mowery 1988, Lewis 1990), but between a British and Chinese company.

Table 9.1   Joint ventures in the Soviet Union and Hungary

	1987	1988	1989	1990	1991(est)
Soviet Union	23	193	1,269	1,754	3,000
Hungary		227	1,357	5,171	

Source: *Financial Times* 20 September 1991; Inzelt and Vincze (1991)

CASE STUDY: RICARDO/DLW

Ricardo Consulting Engineers is a part of Ricardo International plc, an engineering services group formed by the merger in 1990 of Ricardo Group and SAC International. Ricardo has had seventy-five years experience in research, design and development in internal combustion engines. This has been supplemented in the last ten years with a number of UK and international acquisitions to provide it with capabilities in transmission, vehicle and other engineering research and development work. SAC is an international engineering services organization headquartered in Bristol. Its core businesses are in aerospace, nuclear power and manufacturing systems.

Ricardo Group's client list is diverse and the work it undertakes is equally varied. Contracts might range from a total and independent concept-to-production project down to the assessment of a reliability problem on a small component already in series production.

Ricardo International reported turnover of £33m in 1990; SAC contributed £14m to this figure. The Group employs 1700 people with a further 120 sub-contract staff. There are approximately 400 people at the site where Ricardo Consulting Engineers is based.

Over the last ten years, three joint-projects have been undertaken by Ricardo and a Chinese company, Dalian Locomotive and Rolling Stock Works (DLW). The third of these collaborations is still current and has, in 1991, two more years to run. The aims of the projects have been evolutionary and entail the modernization of a Chinese locomotive engine by increasing its power while reducing fuel consumption and improving durability.

The first project was an assessment of the DLW engine. The second cooperative programme began to carry out major improvements to the engine. The third and current collaboration is intended to achieve a further increase in power and performance through a more far-reaching redesign of major components.

DLW is one of three principal locomotive manufacturers in China. The company employs around 12,000 people in the large, cosmopolitan sea port of Dalian on the East Coast of China. There have been engineering works at this site since the late 1880s and, according to Ricardo, DLW has excellent technological and engineering skills.

For Ricardo, the rationale for the collaboration was commercial, as it is first a customer–supplier relationship. DLW, however, looked to collaborate with Ricardo on the engine modernization programme rather than simply commission a turnkey upgrade. It was part of the contract that both parties should work together using the most modern technologies and computational methods. It was also agreed that there would be a process of technology transfer to the Chinese, in the form of intellectual property, tools, equipment and engineering methods.

While DLW looked to the collaboration as a means of achieving an independent and possibly even international locomotive engine manufacturing capability, Ricardo looked upon the relationship as a learning process that would give it access to the world's largest locomotive engine market. Within the company it is believed that it has established a good reputation in the Chinese rail and manufacturing industries and the company is already bidding for other Chinese contracts. It is believed within the company that the likelihood of Ricardo building a substantial engine business in China has been greatly enhanced by their experiences of the three DLW–Ricardo collaborations.

Ricardo has been active in China for some years, although not in the railway industry. During a visit to China in 1978, the then Managing Director of Ricardo was invited by the Chinese Ministry of Railways to give a presentation to a collection of industrialists and bureaucrats. A number of DLW representatives were in the audience and subsequently asked the Managing Director to make a presentation about Ricardo's large engine capabilities in Dalian.

DLW had begun building large, medium-speed diesel engines in the 1960s, based on a Soviet copy of an antiquated USA rail engine. The breakdown in Sino–Soviet communications meant that DLW lost access to Soviet engineers, and at the same time it lost the blueprints for the locomotive engine. DLW responded to this by designing its own engine, largely by imitation. At the time of the first DLW–Ricardo contract, DLW was building 100 of these engines each year. They were, however, very unreliable and had generally poor performance.

The collaboration arose following pressure from the Chinese Ministry of Railways for China's three principal locomotive manufacturers to increase massively the production of diesel engines as steam engines were phased out. The pressure to build more diesel engines was intensified by economic growth and a general increase in the demand for transport. An interim solution was the importation of some 400 General Electric locomotive engines from the USA. The longer-term solution was for Chinese industry to establish the capability to build more and better engines.

DLW felt that an uprating of its existing engine would be the speediest and least costly solution to the demands from the Ministry of Railways. DLW asked Ricardo if it would be prepared to help them in this matter. Ricardo argued that, at that stage, it could only carry out an assessment of the engine to identify problems and, more importantly, the feasibility and likely cost of modernizing the design. DLW had certain cost and performance parameters which could easily have rendered the re-design unworkable.

Ricardo experienced its first major problem with working in a radically different business environment. DLW did not wish to split the work into two phases (assessment and re-design) because it would involve two negotiations with the Ministry of Railways in order to gain acceptance for the project. This was a costly, time-consuming and highly uncertain process. However, Ricardo insisted on the need for an evaluation programme, and this was eventually accepted.

Initially, both parties collaborated in a thorough analysis of the engine. Its design and performance were examined at service load and with a view to its potential uprating. The components for which design improvements would be necessary were identified and recommendations were made for possible methods of improvement. A 20 per cent improvement in power and similar improvements in fuel consumption and durability were targeted. These targets were to be achieved primarily through the addition of a turbocharger and improvements to fuel injection and bearing systems. A full functional specification for an improved design was established as a central working document for the improvement programme.

In 1982, the second programme began. Before confirming the detailed performance targets of the improved engine, both parties entered into an extended negotiation of programme objectives and parameters which would ensure the complementarity of Ricardo's design efforts with DLW's technological and financial resources.

During the project detailed analysis work was carried out using many computer-aided techniques and definitive designs were produced. This was supported by engine test work and an engine was shipped to the UK for performance development work. DLW used substantial foreign exchange to invest in a Data General computer-aided engineering system. This system mirrors that in use at Ricardo and was purchased to facilitate concurrent engineering through the sharing of information.

Fuel injection and turbocharger matching work was carried out by Ricardo and additional single cylinder testing for combustion and fuel injection development were carried out at DLW. DLW manufactured most of the new components; other finished parts were procured from European manufacturers. Rig testing was performed by both parties and DLW then built a prototype for test in a new modern shop at Dalian. Performance and durability tests have been carried out there.

The project was successful and an improved engine was produced, within strict design and manufacturing constraints, at a substantially lower cost and in a shorter time than the production of a totally new engine. The uprated engine has completed over 3,000 hours of successful trials and the improved design is expected to go into full production.

The third and current project will be followed by two more years testing and tooling up. Each of these projects has been a complete contract in itself with no guarantee on either side of future collaboration.

Ricardo rarely enters into collaborative ventures with competitors, either in the UK or internationally. It is particularly concerned that collaboration could lead to the possible loss of markets to partners who were nearly always going to be potential competitors. However, within the company's business there is trend away from turnkey projects and increasingly work involves the full participation of the client in the design and development sequence, accompanied by a conscious interchange of technologies and know-how. This has been the approach taken with DLW.

The management and project team structures in Ricardo have remained the same throughout the collaborations. The Ricardo project teams, for each contract, has consisted of a core of two full time engineers and two or three part time engineers. As and when required, the project team accesses input from other Ricardo functions, such as Design and Contracts. The make-up of the teams varied across the life of the project as it moved from the conceptual to the design and to the analytical phases.

The Ricardo project manager has always been a principal engineer to whom project engineers and development engineers reported. The project manager reports to the manager of the Large Engines Department, who in turn consults with the Technical Director and Managing Director. While the latter three people were only occasionally involved in the operation of the project, all of them played a role at different stages and levels of contract negotiation. In addition to the Ricardo staff, there are typically three or four DLW engineers working with the Ricardo project team at any one time. These are rotated every three or four months.

Ricardo has found organizations in the Far East more wary of entering into international arrangements than their Western counterparts. Building trust and a working relationship at senior management and ministry level had proven even more problematic and time-consuming in the communist Far East. The negotiations over all three contracts have each taken around twelve months to conclude and involved three meetings (two in China, one in UK) with extensive correspondence between both parties. The final meeting of the latest negotiations took five days of solid (ten hour) talks.

Good communications have been vital to the success of the projects. To achieve the objectives a programme was established with regular meetings both in the UK and China to monitor progress. Each partner carried out a number of appropriate tasks and reviewed the work of the other. Particularly important for communication have been the efforts made to achieve a degree of continuity of personnel at both organizations across the three projects. The company has been very careful not to allow changes in personnel to compromise the hard-won working relationship between the two teams. A Ricardo manager recalls how much better the job became once they had broken down the formal barriers and begun to communicate in a more open manner.

The manager of the Large Engines Department has overseen all three stages of the collaboration. However, there have been personnel changes amongst the project managers on Ricardo's side and these have been carefully managed. During the course of the second contract, the first project manager was promoted. He had commanded a great deal of respect at DLW, and he personally introduced the new coordinator to the Chinese. There was concern to maintain the trust between the two parties and the project momentum. To this end a great deal of time was spent with the new project manager until the latter had built his own rapport with the DLW team.

During the third contract, Ricardo decided that it would like to move its manager to take over another project. Over a period of months the second project manager took an increasingly lower profile, leaving one of the other principal engineers to take responsibility for operational coordination. DLW became accustomed to liaising with this third person and after some months Ricardo advised DLW that it would like to make the transition in project management. The hand-over to the third project manager was unproblematic.

Communications were originally hampered by language difficulties. Translators often misinterpreted what was said. Since that time, DLW has sent large numbers of engineers and managers on English courses. In 1990 nearly all of the forty-strong project team spoke English. All of the Ricardo project team members have attempted to learn some words in Chinese.

Communications problems have been exacerbated by the cultural as well as the linguistic divide. It took the Ricardo project team many years of conscious effort to get to grips with Chinese social mores and the communist system. They realized that proficiency in both was a prerequisite for conducting business successfully. The regular meetings between Ricardo and DLW project teams in both China and the UK last two weeks and are, according to Ricardo managers, seen very much as a social as well as an operational review and problem-solving forum. Ricardo has learnt a great deal about Chinese etiquette and preferences, from seating arrangements at restaurants to more personal entertainment and exchanges of gifts.

Ricardo and DLW have both undergone considerable learning at the level of the organization as well as at the level of individual engineers and project teams. As a result of the DLW contracts, it is believed within Ricardo that the whole organization has become far more adept in its dealings with the Far East generally and China in particular.

According to Ricardo staff, DLW is undergoing a difficult process of education in the relevance and use of modern engineering techniques and methods. Its managers and engineers have had to make the enormous transition from the traditional approach of manual estimation, drawing, prototyping, re-drawing and retesting to computational analyses such as computer-aided design and computer-aided testing. Within Ricardo it is believed that the collaboration has assisted DLW engineers' judgements in the appropriate uses of advanced techniques such as Finite Element Analysis and Computational Fluid Dynamics.

Ricardo has not gone out of its way to help DLW build such computational competences. The clause in each of the three contracts referring to technology transfer was intentionally unspecific. Ricardo was happy for DLW to establish an independent engine design capability through the collaboration; it was, however, unwilling to give DLW unconditional access to its proprietary systems.

The first project was characterized by a certain amount of mistrust between the two parties. For example, when Ricardo engineers suggested that a particular action was or was not necessary DLW engineers would approach at least two other Ricardo people to establish that this was indeed the consensus. This resulted in an inordinate amount of extra work for Ricardo staff and generated resentment. The problem was overcome through communication and time. The Ricardo team gradually came to appreciate that the Chinese wanted comprehensive and definitive information to make decisions about any and every course of action. From Ricardo's perspective, the DLW engineers were technically excellent but lacked judgement. For example, while Ricardo engineers might carry out five tests over an afternoon to establish that exhaust gas temperatures remained well within safe working parameters across the engine's full operating range, DLW engineers would have preferred to take several days to plot every possible temperature to several decimal places.

There was another, possibly more important, source of distrust. DLW had not worked with Europeans previously and was understandably concerned that its scarce foreign exchange would be quickly consumed.

Contractual interpretation has been problematic. For example, Ricardo established a number of technical targets as a means to establish the functional specification of the new engine. Target in English implies something that is aimed at, whereas in Chinese, it is an absolute endpoint. So when Ricardo cheerily revealed that it had achieved 97 per cent across all of the target perfromance criteria, DLW considered this to be a shortfall and the Ministry of Railways argued that it was a contractual failure.

Another problem concerned the failure of either party to deliver inputs necessary for the other party's work. Ricardo had experienced a number of delays caused by DLW being late in the delivery of certain components. These problems are difficult to anticipate and by their very nature it is difficult to assess the implications or the appropriate response. According to the manager of Large Engines, the solution is a combination of trust and communication leading to an openness in partners' working relationships with one another. This openness has become much easier over the years for both sides.

The following are identified by the manager of the Large Engines Department as important contributors to the success of the relationship with DLW:

(a) Establishing trust was believed to be a must for such projects to have any hope at all of working.
(b) Establishing mutual respect was also thought to be a prerequisite for a successful working relationship between two such different partners.
(c) Both parties have to be fully committed to the commonly agreed objectives.
(d) Successful contracts benefit everyone.
(e) It is vital that a company understands the idiosyncracies of national negotiating processes.
(f) The active involvement of senior management is critical to the establishment of mutual trust and recognition.
(g) Good project management is essential. A good project manager is likely to be familiar with an industry rather than particular engineering techniques. Project management is all about being in touch, anticipation, encouragement, information dissemination and decisiveness.
(h) It is vital to stick to the brief: do what you are contracted to do. There are often pressures within collaborations to alter priorities, and these can potentially result in missed objectives, time overruns and tension between senior managements. Any decisions to change direction must be dealt with at senior management level and by mutual consent.

CONCLUSIONS

In earlier chapters it was argued that collaboration can encourage learning in firms and supplement their technological capabilities. This chapter has suggested that an international dimension can be added to this learning. There is currently a wide divergence in views on the scale and scope of international technological activities. The existing evidence suggests that technological collaboration is limited to the triad nations, that the majority of large firm R&D activities are domestic, and that international collaboration is primarily unrelated to development activities. This evidence is, however, static and the world is rapidly changing. At present the arguments of Hu (1992) about the lack of internationalization in large companies are compelling. However, there is some evidence that their R&D operations are becoming increasingly international. It is suggested that internationalization extends firms' learning. In the Anglo-Chinese case study, some of the factors found to be important in the other case studies, such as good communications and trust, were found to be even more significant. This highlights the very considerable difficulties in managing international collaborations, particularly with inexperienced partners. The case study also revealed the extent of the learning required amongst partners. This was not only technological, although that was important, but learning about national and business cultures and organization and about differences in the working practices and expectations of individual engineers and managers was also necessary. Although time-consuming, the development of these capabilities provided the basis for the case study company to build its business in a new and very large market.

Chapter 10

Technological collaboration in Japan

Collaboration in technology is believed by many to be qualitatively and quant-itatively different in Japan than it is in other leading industrialized nations. It is a major feature of the strategies of many firms, and is actively promoted by public policies. This chapter discusses some of the features of Japanese industry that encourage collaboration. It suggests that while vertical collaboration is common, and is characterized by close integration of firms enjoying high levels of trust, horizontal collaboration is typified by many of the tensions between cooperative-ness and competition described in earlier chapters. Public policy plays an important role in stimulating horizontal collaborations and policing the tensions within them.

COLLABORATION IN JAPAN

According to one analysis:

> Inter-firm research collaboration in Japan is widespread, structurally diverse, and increasing rapidly in both absolute terms and as a proportion of total R&D... Collaborative research has become the defining feature of Japanese research practice and the *sine qua non* for competitiveness in many technology-intensive sectors.
>
> (Levy and Samuels 1991: 120)

Some of the most commonly referred to attributes of Japanese industry and differences between Japan and the rest of the world are:

1 There are structural reasons explaining a high propensity to collaborate in Japan. As seen in Chapter 1, Japanese industry contributes a greater proportion of gross national expenditure on R&D. As the government and universities undertake less basic and fundamental research than in the West, collaboration between firms may be a proportionately more important activity for firms wishing to access such specialist research.
2 The characteristics of Japanese business increase the propensity to collaborate. The form of business organization – in large groups, or *Keiretsu* – is believed to encourage extensive technological and other linkages between their con-

stituent firms. There is a long tradition of intricate and intimate linkages between subcontractors and their contractors. The employment and training systems of Japanese firms increase the pressures to collaborate.

3 The government has played an important role in stimulating horizontal collaboration, and directing it towards technologies and markets of national importance.

4 There is a cultural and business ethic which elevates the virtues of 'cooperation' over those of competition.

5 Japanese industrial success has depended to a significant extent on the above factors.

As each of these issues is examined in turn, it soon becomes apparent that while there is an element of truth in each they are also based on oversimplification.

Industrial interlinkages

Cooperative research is common in Japan. There are, for example, over fifty collaborative engineering research associations (ERAs) working in a range of technologies. Heaton (1988) describes a 1985 survey by the Japan Fair Trading Commission which found that 55 per cent of 250 leading firms participated in cooperative research. Levy and Samuels (1991) describe a 1986 survey by the Japan Key Technology Centre showing two-thirds of a sample of 261 firms engaged in inter-firm research projects. Aoki (1988) cites a survey by the Agency for Science and Technology which predicted a doubling of the number of collaborating firms during the 1980s, rising to a quarter of all firms by 1990. Leading Japanese electronic firms are sometimes members of a dozen or more engineering research associations (Kodama, quoted in Freeman 1991), and in the computer and aircraft industries the same four or five large, leading firms have been involved in almost every research consortium during the past twenty-five years (Levy and Samuels 1991).

There are, however, a number of misconceptions about horizontal collaboration in Japan. Heaton (1988) refers to these as: the belief that it is the norm in Japanese industry, that it is a government-directed system, and that it is the major means by which Japanese industries achieve technological advances. He argues:

> Once we remove our rose-colored glasses, we see not a formidable machine churning out marketable technology but a modest system of interaction and communication among firms and between the private and public sectors. Instead of coherence and consensus, we see friction and competition that limit the goals of joint research. And instead of fully formulated solutions, we see a system in transition, with the Japanese government struggling to update its policies and redefine its role.
>
> (Heaton 1988: 32)

The belief that Japanese firms collaborate more freely than do Western firms is severely questioned by Fransman's (1990) study of collaboration in Japanese industrial electronics. He found only two examples of 'spontaneous' (i.e. not government promoted) collaboration formed post-war, and argues that Japanese corporations are just as cautious as their Western counterparts in entering cooperative research agreements, and go to great lengths to prevent the leakage of knowledge to competing companies.

One of the major mechanisms which is used in Japanese industry to encourage collaboration is the ERA, and their role has been examined with considerable interest in the West (somewhat paradoxically, as they are replications of the old UK Research Associations; see Chapter 6). The primary work of the associations is, according to Heaton (1988) information exchange and mutual coordination of a small research agenda. The research is almost always conducted in member firms' facilities, and is usually longer-term and more basic than companies typically undertake themselves. He argues that firms rarely join joint R&D projects to obtain financial subsidies, but do so in order to:

(a) gain efficiencies through the solution of common problems;
(b) avoid missing out on important technology or information;
(c) participate in public–private sector consultation;
(d) access intelligence on research in other companies and in universities;
(e) get a critique of their internal technical work from peers in the industry;
(f) obtain a low-cost route to gaining expertise in emerging and cross-disciplinary new fields, enabling faster diversification.

Stenberg (1990), in his study of the development of molecular beam epitaxy (MBE) technology in Japan, argues that there are a variety of reasons why Japanese firms undertake basic R&D. These include the need to have in-house competences in order to identify and correctly evaluate breakthroughs in important new technologies, and also to build up the 'social network relations' with experts in the field to assist this identification and evaluation. Apart from the utility of specific research findings,

> The positive image created from pursuing research at the scientific frontier is seen as a powerful means of building confidence in the company among customers as well as among employees and also makes it easier to attract new qualified scientists and engineers. For image creation it is probably more important that a field can be associated with visionary thinking that sets one company apart from the mainstream than that it shows a large commercial potential.
>
> (Stenberg 1990: 32)

Collaboration also plays an important role in technological diversification. Aoki (1988) describes a survey by the Agency for Science and Technology which found that joint research projects amongst firms in different industries with different capital bases are common and are on the increase. The survey found that the main reason for collaboration according to a majority of respondents was 'to complement and exploit what is lacking and/or heterogenous to its own technology, human resources, know-how, research facility and equipment, etc.' (Aoki 1988: 250).

Business organization

Japanese industry possesses a number of huge business groups – the *Keiretsu* – consisting of a broad range of industrial and financial firms. These groups, such as Mitsui, Mitsubishi and Sumitomo control a diversified range of interests and are tied together by mutual share holdings and relationships with particular main banks. There is an assumption that close trading relationships and cooperation occur within *Keiretsu* (Freeman 1987; Porter 1990). Imai (1988) argues, however, that there has been a gradual loosening of inter-corporate linkages and a blurring of corporate boundaries. He argues that 'Japan's big business groups are not closed groups that make tie-up contracts exclusively within their groups. They contract even with outside firms in other rival groups and most contracts are with firms outside big business groups' (Imai 1990: 170).

In new industrial sectors, such as computers and software, Imai argues, inter-firm linkages have no special connections, or only very weak ones, with traditional business groups. Even in the car industry where suppliers are believed to link very closely with individual assemblers, some firms are supplying a number of competing car manufacturers (Lamming 1992).

Levy and Samuels (1991) argue that most inter-firm research is undertaken by unaffiliated firms, especially in the most innovative sectors. They reveal that of 200,000 cases of joint patent applications by twenty-nine different firms, most were among otherwise unrelated partners.

It seems, therefore, that the *Keiretsu* structures do not inevitably lead to intra-group collaboration. The major strength of the business groups, particularly from the perspectives of firms operating in the UK and USA financial systems, is the way long-term relationships and continuing information exchange occur between the banks and industrial companies in the group, thus making longer-term, and perhaps collaborative, research more acceptable.

The Japanese employment and training systems have an impact on collaborative activity. Large Japanese corporations operate a life-time employment system for core employees. In order to continue to meet the increasing costs of this system (through limited inflation, growing expectations from workers and the increasing age, and hence salaries, of employees), companies need continually to grow. This often requires diversification. A number of Japanese firms are increasingly making acquisitions to assist this diversification; the purchases of USA biotechnology firms are a case in point. Another important element in their diversification strategies is collaboration. Prahalad and Hamel (1990), for example, argue that NEC foresaw the convergence of semiconductors, telecommunications and computer businesses and established a strong position in each of these areas through the assistance of a 'myriad' of strategic alliances, including over 100 in semiconductors.

Japanese firms tend to train their scientists and engineers on-the-job in ways that extend the development of in-house capabilities and specialist, firm-specific knowledge in established technologies and business areas. Their employment systems involve recruitment of young people and their integration into an internal labour market where careers and promotion depend on comparisons of performance with fellow employees. As Aoki and Rosenberg (1987) argue, these employment system characteristics provide a problem for firms faced with diversification and interdisciplinary research, as they are

> less flexible in gathering specialists from different disciplines as the agenda for research and problem-solving requires. Recruiting ready-made specialists in needed disciplines would be in conflict with the basic imperatives of the ranking hierarchy as an incentive device: to employ engineers and researchers out of schools and promote them within the ranking hierarchy based on long-term in-house competition.
>
> (Aoki and Rosenberg 1987: 20)

Collaboration can be a mechanism for overcoming the rigidities of the Japanese employment systems in this respect.

The vertical technological linkages between contractors and sub-contractors in Japan are argued to be a major source of technological dynamism (Economic Planning Agency 1990). These relationships have evolved considerably since the 1950s when sub-contractors were seen as low-cost suppliers which were dispensable in times of recession. Sako (1991) cites a study by the Japanese Fair Trade Commission showing that the majority of large firms' relationships with their suppliers were long-term and continuous, and that price considerations are less important than the stability of good quality products. One of the reasons for the change in these relationships was a concern on the part of the large firms, and of the government, to extend skills throughout industry. Freeman (1991) suggests that there are also innovative efficiency reasons.

> Modification or innovation of a part or component of a product or process by one sub-contractor inevitably affected the manufacturing process of the whole. Especially in electronics, innovation among sub-contractors is subject to the constraint of compatibility with the customer's (or parent company's) technology. Therefore, the sub-contractor must supply a product according to detailed specifications which can only be modified within certain limits. To a degree this compels parent companies to offer advice and supply the necessary technology to subcontractors so as to increase their economic and engineering capabilities. This results in a higher dependence on up-graded sub-contractors because of their specialized technology and equipment instead of the traditional low cost approach... As the technical competence of sub-contractors improved, a more equal relationship between large and small enterprises began to develop in many cases.
>
> (Freeman 1991: 505)

Sako (1992) distinguishes two ideal types of buyer–supplier relationships at either end of a continuum of actual practices: Arm's-length Contractual Relations (ACR) and Obligational Contractual Relations (OCR).

> At one extreme, firms rely on ACR if they wish to retain full control over one's destiny. Independence is the guiding principle here, which involves not only being unaffected by the decisions of other companies, but also by one's own decisions (e.g. over sourcing and sales) made in the past. This often requires not disclosing much information (e.g. about costing and future plans) to existing and potential buyers and suppliers. The arm's length nature of contracts enables firms to engage in a hard commercial bargain to obtain competitive prices, although an excessive use of threats and bluffs may make some firms wary of too much antagonism. At the other extreme, firms enter into an OCR if they prefer high trust cooperativeness with a commitment to trade over the long run. This commitment may come at the expense of taking on rather a lot of sometimes onerous obligations and requests (e.g. for just-in-time and ship-to-stock delivery). But the benefits of accepting mutual obligations lie in good quality and service, growing or stable orders and other non-price aspects of trading born out of a tacit understanding over time.
>
> (Sako 1992: 2)

In her study of subcontracting relationships in Japan and Britain, Sako found that OCR has distinct advantages for prices and quality. OCR is dependent on high levels of trust, and that it is this in the context of the history of the development of institutional arrangements in industry which makes it more common in Japan, and that much more difficult to export to other countries like Britain which are trying to move away from ACR to OCR practices.

Lamming's (1992) analysis of the 'lean supply' relationship in the automotive industry argues that in conjunction with 'lean production' (Womack *et al*. 1991) supplier firms need increasingly to provide their own technology. Supplier firms in Japan have traditionally played an important role:

> The ability of the Japanese firms to operate efficiently while using a larger fraction of unique parts is due in significant part to the capability of the supplier network. The implication is that it is not only the extent of supplier involvement that is important but the quality of the relationship and the way that it is managed that matters... It is important to note that such benefits are based on a relationship of reciprocity. Not only do suppliers have valuable capability, but the auto firms managesd the process so that capability plays an important role. Moreover, the auto firms cultivate capability in their suppliers.
>
> (Clark 1989: 1256)

Sub-contracting relationships in Japan generally involve linkages between large and small firms, and while small firms are involved in the new product development activities of the large firms they are believed to undertake little R&D themselves. However, small firms are argued by some to be increasing their own R&D activities. According to Furukawa, Teramoto and Kanda (1990) there has been a growth in the number of small firms collaborating together to overcome the costs of undertaking R&D. They describe groups of small firms – 'Inter-industrial Networks for Technological Activities' – working together to create sufficient resources to undertake R&D and to better utilize existing, accumulated technology. They cite a survey showing that the number of such networks increased from 357 in 1984 to 981 in 1986, and they argue that many of these groups have allowed small firms to strengthen their R&D and develop new products.

Government policy

It has been argued that the policies adopted by the government, in particular the Ministry of International Trade and Industry (MITI), have played an important role in the development of Japanese technological competitiveness (Freeman 1987).

> The guiding hand of MITI (has) had a considerable influence in shaping the long-term pattern of structural change in the Japanese economy and this influence was exerted largely on the basis of judgements about the future direction of technical change and the relative importance of various technologies.
>
> (Freeman 1987: 34)

In his analysis of the 'Japanese Technology-Creating System', Fransman (1990) argues that the government has played an important, but changing, role in its formation. Thus, following the war, it was government research laboratories and universities that took the lead in acquiring and diffusing new information technology. From 1960 onwards, in-house R&D capacities that had been built up in firms replaced government laboratories in importance, but the government remained highly active in initiating cooperative research programmes.

> MITI officials were aware of the potential benefits of interfirm cooperative research. These benefits could be particularly significant in areas of oriented basic research which do not result in immediately commercializable outputs and where, accordingly, competing firms might be expected to find common ground which could serve as the basis for research cooperation.
>
> (Fransman 1990: 9)

Stenberg (1990) describes how valuable was the support of MITI in encouraging R&D activities in MBE in Japanese industry. MITI launched three research programmes in the area early in the development of the technology, and its Electrotechnical Laboratory (MITI's main electronics labratory) in addition to planning and monitoring these programmes, also undertook research in the area itself.

MITI has not, of course, been infallible, and some of its judgements have proven to be wrong. In the car industry, for example, at one time MITI was impressed by the USA's 'big three', and was keen to reduce Japan's twelve car manufacturers to emulate the US system. At another time it tried to press Honda not to produce cars.

Japanese ministries possess a number of agencies developed to encourage collaboration and technology transfer. Levy and Samuels (1991) contend that the 1980s witnessed a 'veritable explosion' of new forms of government-backed joint research, including: informal 'forums' which precede the formation of ERAs; research cartels in certain industries; programmes to enhance the access of research consortia to government laboratories; and the Japanese Key Technology Centre (Key-TEC). This independent corporation with its own research facilities was jointly established by MITI and the Ministry of Posts and Telecommunications. These organizations provide 70 per cent of funds, while private funding, from companies such as Mitsubishi Chemical, Fujitsu, Toshiba and NEC, provides the remaining 30 per cent. According to Heaton (1988):

> In funding joint research, Key-TEC eschews targeting specific sectors, prefer-ring instead to support 'industrial and communications technology' in whichever areas the most promising proposals appear. Rather than issuing grants, contracts, or loans, Key-TEC typically purchases stock in new ventures, with the intention of later selling it at a profit. To receive funding, a venture must be set up by at least two pre-existing enterprises. A typical Key-TEC venture is the Protein Engineering Research Institute, established in 1987 by 15 diversified chemical companies. Its budget is \$150 million (70 per cent of it from Key-TEC) over 10 years... Its agenda – to perform basic research as a precursor to commerciable technologies – was established solely by the share-holders.
>
> (Heaton 1988: 35)

Levy and Samuels (1991) suggest that Key-TEC does additionally make loans to joint-venture research firms, and also has a role in collecting and diffusing scientific and technical information and promoting international and domestic research cooperation. Other projects Key-TEC is working on include research on electronic dictionaries and a number of media technologies.

Fransman and Tanaka (1991) have evaluated the role of the Protein Engineering Research Institute (PERI), which is considered one of the great successes of Key-TEC. In its organization, they argue, it resembles another MITI-supported group: the Institute for New Generation Computer Technology (ICOT). Its purpose is to undertake generic research into protein engineering. The more basic research is undertaken in central PERI laboratories, while the more applications-oriented research is undertaken in individual member companies.

> Typically, the researchers from the member companies spend about three years in PERI after which they rotate back to their companies and are replaced by other company researchers. At times parallel research groups are established inside the companies whose research closely follows that being done in PERI, but with concentration on applications. In this way there is a relatively smooth transfer of generic knowledge from PERI, where joint cooperative research is

undertaken, to the member companies where private applications-oriented research is done.

(Fransman and Tanaka 1991: 21)

The strengths of PERI, according to Fransman and Tanaka, are:

1 like similar organizations such as ICOT, it is able to tap effectively the world stock of knowledge, assisted by its high international profile;
2 it has provided an effective method of interaction with Japanese universities (such university/industry interaction has proven difficult in the past);
3 it has subsidized the development and diffusion of new generic technology: 'Although ... it is likely that the companies would in any event, even without this or similar projects, have acquired protein engineering capabilities, it is probable that they have done so faster and more cost-effectively as a result of this project' (Fransman and Tanaka 1991: 22);
4 it has avoided research overlap in individual companies;
5 the distinctive competences of researchers in individual companies have been combined;
6 expensive, skill-intensive equipment has been shared;
7 it has increased the diffusion of generic protein engineering technology;
8 it has focussed national attention on protein engineering as a strategic technology;
9 it has assisted the transfer of inter-disciplinary and tacit knowledge between companies.

Fransman (1990) argues that while the forms of organization that give Japanese industry so much advantage in innovation have been created without government influence, government has helped in a number of ways through its collaboration-promoting activities. The government and major procurers, such as NTT, have reduced the uncertainties, and the transaction costs, of establishing research co-operation, and thus helped companies benefit from it. The research associations are mainly financed by public funds, and are also promoted through favourable tax and regulatory policy (Heaton, 1988). The government has also encouraged a broader franchise of participants in collaboration, thus assisting technological diffusion.

Keen observers of the role the Japanese government plays in promoting collaboration emphasize the elaborate political bargaining process which typifies the formation and functioning of a collaborative programme. Levy and Samuels (1991) refer to the prolonged process of conflict and negotiation in establishing programmes. They suggest a typical process involving MITI proposing a joint research venture (on industry's or its own initiative); the proposal then being attacked by firms for being too far from market, or for excluding them, by the Ministry of Finance for being unnecessary or too costly, and by rival ministries for encroaching on their territory (and which then set up rival programmes). For an example of the machinations involved in setting up a collaborative programme, readers are referred to Levy and Samuels' account of the tortuous formation of Key-TEC.

Culture and business ethos

Japanese culture is said to be based upon the Confucian ethic of original virtue, rather than the Catholic view of original sin (Dore 1986). Certainly, individual, business and government behaviour is much more directed towards the attainment of cooperation and harmony than occurs in the West, and a high level of trust is a cultural disposition in Japan and this underpins business relationships (Dore 1973). However, the competitive tensions between firms in Japan frequently erupt in noncooperative behaviour. Levy and Samuels (1991) argue that even in circumstances where the logic of cooperation is particularly strong, cooperative behaviour is not assured without the presence of government authority to reward cooperation and occasionally punish defection (through exclusion in subsequent projects). They suggest:

> The instances of non-cooperative behaviour – lead firms in an industry refusing to join a consortium, information hoarding, lackadaisical research efforts – are too numerous to describe... In short, the oligopolistic structure of Japanese industry facilitates cooperation in the context of active state support, but it is insufficient to generate inter-firm collaboration by itself... Japanese firms, ever distrustful of their competitors, have by no means cast away all their fears and embraced full-scale cooperation. Rather, they have developed a set of practices which allow them to pool resources while keeping their partners at arm's length. Typically, researchers from different companies do not work together on the same technical problem. Instead, each participant company assumes responsibility for a specific task. Research is performed independently in each firm's own labs, with the patents then shared. This practice, referred to as 'distributed cooperation' by the Japanese, suggests an exchange of roughly comparable, independently-produced technologies rather than genuine collaboration.
>
> (Levy and Samuels 1991: 128–9)

At the same time, groups of cooperating firms and government agencies are argued to develop strong integrating bonds. Ouchi (1986) describes a 'social memory' comprising groups of Japanese firms and government agencies that develop community senses of social responsibility and equity. These groups remember past cooperativeness and conflict and have a capacity for reward and punishment.

Industrial success

Collaboration in Japan takes a variety of forms and affects different industries and different sized firms in various ways at different times. Given this complexity it is difficult to directly ascribe collaboration as a major contributor to Japan's technological and economic success. Nevertheless, Fransman (1990) in his detailed historical study argues that research cooperation has brought significant benefits to the Japanese industrial electronics industry. By referring to the close sub-contracting relationships in the automobile and electronics industry in Japan, Sako

(1991) argues that it has become the accepted view that such long-term, continuous buyer–supplier transactions account for the international competitiveness of manufactured goods in Japan. Imai (1990) contends:

> it is clear that Japan has built up an efficient industrial system that creates innovation through interactive coordination between production, marketing and R&D. Dense information-sharing and communication across the boundaries of firms and industries have been keys to success.
>
> (Imai 1990: 184)

CONCLUSIONS

Virtually all the informed analyses of collaboration in Japan argue the importance of government support. Although originally devised to improve the capacities of SMEs, Japanese horizontal collaborative programmes have become an important means of technology development for large firms. The government has played an important role in creating institutions in which collaborative research can take place, and in creating a tradition of collaborative R&D projects. Although the level of actual horizontal cooperation between firms is open to doubt, there is little argument that the collaborations which do exist provide important mechanisms for information transfer, and hence potentially improved innovation. One of the most important benefits of collaboration is achieved through personnel transfer.

Technological collaboration has been encouraged by consistent public policy. That is, firms enter collaborative programmes in the expectation of future programmes which they may value particularly highly. This encourages cooperative behaviour through the fear of exclusion. Whereas vertical technological collaboration is very common, and occurs spontaneously, horizontal collaboration frequently requires government stimulus. There are implications of this for the process of collaboration, and in the level of trust between partners. Unlike the high-trust vertical collaborations, horizontal collaboration in Japan is typified by the tensions between cooperativeness and competition found in Western economies. While the aims of horizontal collaboration are similar to those in the West, such as learning and building core capabilities and technological diversification, the circumstances in which it occurs are different. This applies not only to the public policy considerations mentioned above, but to the greater levels of cooperativeness generally in Japanese society, and the wider experience of the value of long-term, trust-based vertical relationships.

How far the Japanese system of innovation needs to adapt now that Japan is a leader rather than a follower in many key technologies is very much open to debate, as is the role that collaboration will pay in any future system. There are those who argue that in the future Japanese firms will place much less reliance on externally-acquired technology (Komoda 1989). Nevertheless, the Japanese system is continuing, as it has in the past, to adapt and develop. Imai (1990) argues, however, that there is a need for further economic and social evolution. Amongst the factors which he argues need to be addressed are the need for greater multiplicity of sources of innovation. Japan, for example, could not provide the circumstances which led to the development of biotechnology in the USA. In does not have the same tradition of academic spin-offs, a supportive environment for small high-technology firms, and employee mobility to enhance technology transfer. But Imai also refers to the way 'Current technological conditions are also promoting the reshuffling process of the Japanese industrial system. As is typical in information technologies, new high technologies require new linkages between firms, wholesalers and retailers, and research organizations' (Imai 1990: 185).

As part of the continuing adaptation of the Japanese system of innovation, and the role of innovation within it, there is a growing priority placed by many on the question of internationalization. Japanese industry, it is argued, needs increasingly to access and develop science and technology internationally. The increased foreign investment by Japanese firms in R&D laboratories and in high-technology acquisitions referred to earlier are indications of this development.

Chapter 11

The collaborative technological activities of small firms

This chapter argues that many innovative smaller firms possess advantages over larger firms and can, often in combination with large firms, contribute importantly to technological development. Innovative small firms enjoy dense and diverse external linkages both vertically and horizontally. These firms learn technologically not only through R&D but also through a variety of manufacturing and marketing linkages. It will be argued that a number of constraints affect small firm linkages. These include: the shortage of key personnel; problems of linking with large firms; and inappropriate public policies.

SMALL FIRM COLLABORATIONS

Both large firms and small and medium-sized enterprises (SMEs) can make important contributions to technological innovation, albeit their relative contributions vary very considerably from sector to sector and can change over the industry life cycle. Table 11.1 shows the large number of small firms in high technology in the USA, and the number of innovations they contribute. Not only do large and small firms *separately* play an important role in technological innovation, but they often play interactive and *complementary* roles. In the case of the emergence of new technology-based sectors (e.g. semiconductors and CAD), Rothwell (1983; 1989a) has described the *dynamic complementarities* existing between large and small firms that played an important role in industrial evolution, and Chapter 5 has described this relationship in biotechnology. In the scientific and medical instruments sectors large informed users can act as primary sources of technology for small suppliers (von Hippel 1976; Shaw 1988). In some areas large firms employ small firms as a 'window' on new technology (Roberts and Berry 1985). And technological exchanges between small suppliers and their larger customers are becoming an increasingly common phenomenon in manufacturing. In general, leading-edge customers can play an important role in 'pulling' innovations from their suppliers, both large and small (Rothwell 1986).

Partnerships between larger and smaller firms have multiplied over the past few years. To large firms, partnerships usually offer a channel to tap into the

Table 11.1a   US small businesses active in some high technology areas, by industry

Industry	Small firms	% of Total firms
Automation	1,566	94
Biotechnology	414	95
Computers	1,896	93
Advanced materials	636	88
Services	4,755	96
Software	5,844	98
Telecomms	1,094	93

Source: NSF (1989)

Table 11.1b   Number of innovations for large and small firms in the most innovative industries

	Large firm innovations	Small firm innovations
Electronic computing equipment	158	227
Process control instruments	68	93
Radio & TV communicating equipment	83	72
Pharmaceutical preparations	120	13
Electronic components	54	73
Engineering & scientific instruments	43	83
Semiconductors	91	29
Plastic products	22	82
Photographic equipment	79	9

Source: Acs and Andretch (1990)

innovative and entrepreneurial potential of smaller companies, and to overcome some of their own rigidities. In most of the observed partnerships, smaller firms perform research and development for the larger firms and/or transfer innovations to them. These larger firms offer their smaller partners the ability to reach world markets quickly, without having to build their own infrastructure or to negotiate complex agreements with multiple agents. Larger firms also offer the experience of volume manufacturing. The complementarity is obvious.

(Doz 1988: 317)

A number of established modes of large firm/small firm interaction are listed in Table 11.2 (Rothwell 1989b) and it can be seen that many of these involve, centrally, inter-firm technological exchanges. Some of the linkages focus on the supplier/manufacturer relationships during product development; others operate at the manufacturer/customer interface and can involve both technological and 'customer-need' informational exchange; others involve mainly financial transactions

Table 11.2 Some modes of large and small firm interaction

Manufacturing sub-contracting relationships

SMFs supply components and sub-assemblies to large companies. As part of this process large companies frequently transfer technological, manufacturing and quality control know-how to their small suppliers. Stable relationships can develop which are mutually advantageous.

Producer/customer relationships

SMFs supply finished products to large companies. Large companies can transfer technological know-how and supply suggestions for improvements to small suppliers based on user experience.

Licensing agreements

Large companies provide licences to small firms for innovative new developments. This frequently involves technology that the large company does not wish to exploit in-house but which it wishes to gain a financial return on. In some cases it can involve technology which the large company will subsequently purchase in the form of equipment for in-house use, for example large process companies transferring new process control technology to small instrument companies.

Contracting-out R&D

Large companies fund targeted R&D in small specialist consultancy companies, e.g. automobile companies funding R&D in specialist engine developers; pharmaceutical companies funding R&D in small biotechnology companies.

Collaborative development

Large and small companies collaborate in the development of a new product for the large company, e.g. small software or design houses collaborating respectively with large computer and automobile manufacturers.

Large/small firm joint ventures

Large and small firms collaborate in the development of an innovative new product containing technology new to the large partner. The large firm provides financial, manufacturing and marketing resources; the small firm provides specialist technological know-how and entrepreneurial dynamism. Generally the new products are complementary to the large firms' product range. They are manufactured by the small partner.

Educational acquisitions

Large companies acquire NTBFs to provide them with a window on new technology and an entree to new business areas. Examples of this are fairly common in the new-wave biotechnology field.

Sponsored spin-outs

The large company offers financial backing for entrepreneurial employees to spin-out to form a new small firm to exploit technology developed within the parent company, but which is deemed unsuitable for in-house exploitation.

Venture nurturing

The large company offers not only financial support to the sponsored spinout, but also access to managerial, marketing and manufacturing expertise and, if appropriate, to channels of distribution.

Independent spin-out assistance

The large company offers technical assistance to an independent spin-out and sometimes acts as first customer for its products. Pre-payments can provide a crucial source of income to the new company.

Personnel secondment

A number of large European companies have developed schemes to loan experienced managers to assist new and existing SMFs in their locality.

Source: Rothwell (1989) 'SMFs, Inter Firm Relationships and Technological Change', in *Entrepreneurship and Regional Development* 1: 275-91 (e.g. sponsored spin-outs and corporate ventures). Some firms focus entirely on technological co-developments; many will involve some combination of these.

In order to understand why large companies see advantages in technological and related interactions with small firms and *vice versa*, it is first necessary to understand large and small firms' relative advantages in innovation, and the most important of these are listed in Table 11.3 (Rothwell 1983). According to this listing, the innovatory advantages and capabilities of large firms are mainly material (large financial and qualified manpower resources; extended external scientific and technological networks; large marketing resources; comprehensive range of management skills, etc.); while those of small firms are mainly behavioural (management dynamism; organizational flexibility; rapid internal communication; high degree of adaptability, etc.). Large/small interactions clearly can assist both partners to overcome their innovatory disadvantages and at least partially combine their respective behavioural and material innovatory advantages and capabilities.

As argued in Chapter 3, it is especially during periods in which the pace of technological change is fast, and when there is great uncertainty around the many new technological and market opportunities, that the failure of firms of all sizes to gain early access to newly emerging capabilities can rapidly result in obsolescence in their products and processes obsolescence. Appropriately constructed strategies, based on a combination of in-house technological accumulation complemented by external inputs can, in contrast, enable firms technologically to update existing products and/or move to new product areas (Dodgson and Rothwell 1987; Mayer-Krahmer and Kuntze 1987). External linkages are an important element of corporate technology strategies directed towards creating a network plugging the firm into appropriate sources of complementary technological information and expertise (Dodgson and Rothwell 1989).

Links between large and small firms are a focus of industry and technology policy in many nations. Increasingly, the EC is attempting to expand large/small firm interaction through its collaborative programmes. One of the major functions of the Commission's Directorate for Small Firms is to advise on how small firms might become more involved in collaborative R&D projects. At the same time, public policy has concentrated on creating technology transfer mechanisms which benefit small firms (Rothwell and Dodgson 1992).

THE DENSE AND DIVERSE EXTERNAL LINKAGES OF SMES

Roberts (1991) shows the importance of strong initial technology transfer links in the formation and subsequent success of high technology SMEs, and innovative SMEs enjoy a continuing high level of linkages with external agencies. This is shown by a questionnaire-based study of 100 innovative SMEs in the UK (Beesley and Rothwell 1987), and a study of eighty high technology firms (almost all of which are small) in Italy (Parolini 1990).

The Beesley and Rothwell study found that overall, 89 per cent of the firms studied had a significant link in at least one of the following areas: contract-out

Table 11.3 Advantages and disadvantages of small and large firms in innovation

	Small firms	Large firms
Marketing	Ability to react quickly to keep abreast of fast changing market requirements. (Market start-up abroad can be prohibitively costly.)	Comprehensive distribution and servicing facilities. High degree to market power with existing products.
Management	Lack of bureaucracy. Dynamic, entrepreneurial managers react quickly to take advantage of new opportunities and are willing to accept risk.	Professional managers able to control complex organizations and establish corporate strategies. Can suffer an excess of bureaucracy. Often controlled by accountants who can be risk-averse. Managers can become mere 'administrators' who lack dynamism with respect to new long-term opportunities.
Internal communication	Efficient and informal internal communication networks. Affords a fast response to internal problem solving; provides ability to reorganize rapidly to adapt to change in the external environment.	Internal communications often cumbersome; this can lead to slow reaction to external threats and opportunities.
Qualified technical manpower	Often lack suitably qualified technical specialists. Often unable to support a formal R&D effort on an appreciable scale.	Ability to attract highly skilled technical specialists. Can support the establishment of a large R&D laboratory.
External communications	Often lack the time or resources to identify and use important external sources of scientific and technological expertise.	Able to plug-in to external sources of scientific and technological expertise. Can afford library and information services. Can subcontract R&D to specialist centres of expertise. Can buy crucial technical information and technology.
Finance	Can experience great difficulty in attracting capital, especially risk capital. Innovation can represent a disproportionately large financial risk. Inability to spread risk over a portfolio of projects.	Ability to borrow on capital market. Ability to spread risk over a portfolio of projects. Better able to fund diversification into new technologies and new markets.
Economies of scale and the systems approach	In some areas scale economies form substantial entry barrier to small firms. Inability to offer integrated product lines or systems.	Ability to gain scale economies in R&D production and marketing. Ability to offer a range of complementary products. Ability to offer a range of complementary products. Ability to bid for large turnkey projects.
Growth	Can experience difficulty in acquiring external capital necessary for rapid growth. Entrepreneurial managers sometimes unable to cope with increasingly complex organizations.	Ability to finance expansion of production base. Ability to fund growth via diversification and acquisition.
Patents	Can experience problems in coping with the patent system. Cannot afford time or costs involved in patent litigation.	Ability to employ patent specialists. Can afford to litigate to defend patents against infringement.
Government regulations	Often cannot cope with complex regulations. Unit costs of compliance for small firms often high.	Ability to fund legal services to cope with complex regulatory requirements. Can spread regulatory costs. Able to fund R&D necessary for compliance.

Source: Rothwell (1983) 'Innovation and Firm Size: A Case for Dynamic Complementarity', *Journal of General Management*, 8, 3, Spring.

R&D; joint R&D ventures; marketing relationships; manufacturing; links with educational establishments, other public sector bodies and research associations. Excluding links with public sector bodies and taking into account only those links with other firms, this figure drops to 84.5 per cent. Looking at only technical and marketing links, 69 per cent of the firms had some form of significant link. Finally, excluding marketing links, 47.5 per cent of the firms had some form of technical link with other firms.

The range of non-technical links were included as, following a previous pilot study, it was found that there was a whole range of information inputs crucial to SME innovation which involved considerably more than a straightforward transfer of technology or a formal joint venture agreement. It was clear from the pilot study that companies used activity in one area to gain access to expertise relating to another of their activities; for example, undertaking agency agreements in order to gain access to technology or to information on markets new to the firm.

The Parolini study also found a high level of linkage, with 63.8 per cent of the firms engaged in agreements with other companies. The types of agreements are shown in Table 11.4. This study again highlights the density of agreements between firms, and the diversity of the SME linkages which include not only technology agreements, but very extensive manufacturing and marketing agreements.

Table 11.4   Kinds of agreements in high technology companies

Kind of agreement	Percentage
Technology transfer	29.4
Joint R&D	15.7
Joint manufacturing	21.6
Other manufacturing agreements	60.8
Marketing agreements	54.9

Source: Parolini (1990)

The Beesley and Rothwell study examined the types of linkages in some depth. These are discussed below, and are taken from Beesley and Rothwell (1987).

Sub-contracted R&D

39 per cent of the firms in the survey contracted out R&D. Of the forty firms involved, 70 per cent contracted out 10 per cent or less of their total R&D; 14 per cent contracted out between 11 and 20 per cent; 2.5 per cent contracted out between 21 and 30 per cent; 85.7 per cent of sub-contracting R&D relationships were with firms employing 200–499; and only 5 per cent were with firms employing more than 500. 47 per cent of the contractees were supplier firms, and 18.5 per cent were customers.

In 34 per cent of the cases the main motive for contracting out R&D was to gain access to technology new to the firm, and in 50 per cent of the cases the motive was to shorten lead times. 53 per cent of the contracting out firms were forced to undertake such action because of a lack of skills.

Collaborative R&D

26 per cent of the firms engaged in some form of collaborative R&D venture. Most of the collaborations (56 per cent) took place with firms employing less than 200. 46.6 per cent of the collaborating firms collaborated with suppliers. 34.6 per cent of the firms engaged in formal R&D ventures. However, a further 65 per cent of the firms took part in collaborative R&D which involved active technical participation, not necessarily on a formal basis. 73 per cent of the firms taking part in collaborative R&D played an active role in initiating the collaboration, with a similar percentage (74 per cent) having a policy of actively seeking external ideas. Clearly the R&D linkage activity in these firms is strategic in nature. The majority of firms with collaborative R&D (61.5 per cent) already had established contacts with their R&D partners prior to the collaboration mentioned.

The above two sets of data indicate a strong pattern of technical interchange and liaison during product development at the component supplier/manufacturer interface.

Marketing relationships

Agency

25 per cent of all the firms in the survey acted as agents for other firms. Of this 25 per cent, most had agency agreements with foreign firms (84 per cent). The major reason for engaging in agency agreements was to gain access to products which complemented the firms' own product range (85 per cent). This was an effective way of making up for an in-house product development deficiency. However, 12 per cent of the firms who acted as agents used the agency to gain access directly to new technological know-how, and 16 per cent used it to gain knowledge of a new market prior to innovating their own products.

Collaborative marketing

37 per cent of all the firms were involved with collaborative marketing arrangements, the major proportion of which (72 per cent) took the form of geographical marketing agreements. 46 per cent of the firms with collaborative marketing had arrangements to sell mutually complementary products. A major reason for engaging in collaborative marketing was to gain access to new sectors of industry (74 per cent), whilst 20 per cent of the firms used the marketing agreements to gain

knowledge of a new market area, and 26 per cent to gain access to new technological know-how.

Manufacturing relationships

Sub-contracting manufacturing

For many SMEs, sub-contracting manufacturing can be an important means of gaining access to new production technologies, and can enable firms to innovate products requiring new production techniques, without having to invest heavily in expensive, sophisticated production equipment. A large proportion of the firms (68 per cent) sub-contracted some of their manufacturing. 52 per cent of these firms sub-contracted less than 10 per cent of their total production, but 20 per cent sub-contracted between 10 and 20 per cent and 27 per cent sub-contracted over 20 per cent.

Licensing

Only 16 per cent of all the firms manufactured other firms' products under licence, accounting for under 20 per cent of total production in 87 per cent of the cases. In 60 per cent of the cases, a major reason for licensing was to gain knowledge of new technological know-how; 34 per cent of the firms were also using licensing as a means of gaining access to new market areas, and 27 per cent to gain knowledge of new markets in order to innovate in-house products in the future.

Public sector linkages

55 per cent of the firms had regular contact with R&D activities being carried out in public sector institutes, and 51 per cent had enjoyed technical inputs from the public sector of use to their innovation activities. 31 per cent of all the firms sponsored students at educational establishments, and 39 per cent provided placements for sandwich students. Sponsorship and placement also provide a useful means of gaining access to state-of-the-art research and knowledge.

Data were also obtained on the take-up of government schemes by the sample of firms. 40 per cent of the firms took up at least one government-funded scheme in order to gain access to external technical expertise. Of these firms, 57 per cent took up only one scheme, 33 per cent took up two schemes, and 10 per cent took up three schemes.

Following the postal questionnaire, a follow-up telephone study of twenty of the surveyed firms reinforced the findings. The majority of managers emphasised their continuing commitment to external linkage activities, stating their belief that they were likely to become an even more significant element in their firms' continued survival and future growth.

External linkages: the human resource factor

From the viewpoint of external technical communication, SMEs often lack the resources to identify and use important external sources of scientific and technological expertise. Acquiring external know-how is not a costless process and, for a SME, it can carry high opportunity costs. Table 11.5 provides some data on the extent of SME linkages with universities in the USA and the way in which size affects the type of relationships established. It can be seen that there is a distinct increase in the various kinds of linkage as the size of company increases.

SMEs can be further disadvantaged in their external search processes (for new information or ideas) and other learning activities, since they often lack suitably qualified technical specialists. Such personnel are often scarce, and therefore too expensive for SMEs, but this problem is often compounded by a lack of awareness on the part of smaller firms of the value of employing highly skilled personnel. It has been suggested that a SME's ability to access external know-how is conditioned by its in-house employment of qualified scientists and technologists (Angell *et al.* 1984). Of equal importance is the fact that lack of qualified scientists and engineers (QSE) can inhibit the SME's ability to disseminate and further develop technological know-how, even when it does succeed in acquiring it from external sources. This was evidenced in a study of barriers to growth on SMEs in the UK:

> There was no evidence in the material surveyed to suggest that there are infrastructural deficiencies in the UK that particularly inhibit SMEs' access to external technology and external expertise. The most important factors determining a SME's propensity and ability to access external sources of technology are internal to the firm, most notably the employment of QSEs and the outwards-lookingness of management. The employment of QSEs is also an important

Table 11.5 Ways of relating to universities reported by business respondents, by size of company

Ways of relating %	Size of company (employees)			
	1–19	20–99	100–499	500+
Employees attended conferences	30	43	59	73
Company recruited graduates	19	41	57	81
Company got technical services	47	35	35	39
Company provided financial aid for courses	17	33	58	75
Informal unpaid interaction	25	23	28	45
Company donated funds/equipment	18	26	34	50
Company employed faculty consultants	14	19	27	40
Research contracts	6	6	10	27
Research grants/consortia	4	3	8	19

Source: NSF (1987)

determinant of a SME's ability effectively to assimilate and exploit bought-in technology.

(Rothwell and Beesley 1989)

A lack of QSE employment does not mean that a SME will be unable to conduct any form of external technological search; it will, however, impose certain limitations on, and might mean a significantly different pattern of, search. For example, a study of innovation in textile machinery conducted some years ago, showed that the twenty-five firms producing 'technologically significant' innovations had a different pattern of external contacts from the fifteen firms producing 'technologically incremental' innovations; and that these different patterns reflected some degree of different internal technical manpower endowments (Rothwell 1976). This study suggested that innovation in those companies producing significant innovations is generally more externally linked than in the incremental innovators; that their external linkage activity is more technically oriented; and that these patterns are linked to the differences in in-house QSE employment. These results tend to confirm the proposition that lack of QSE in a SME generally *doubly* inhibits technological accumulation, through limits on the more obvious mechanism of in-house R&D and through the inability satisfactorily to access external know-how. Clearly such a firm will suffer severe limitations on its range of product development activities. The importance of having employees capable of being receptive to external sources of information applies not only to *product* innovation but also to *process* innovation. In a major study of the diffusion of programmable automation in the USA, Kelley and Brooks (1990) found that the presence of an outward-looking person, well linked to the external technological and commercial environment, considerably increased the propensity to invest in new manufacturing equipment, such as CNC machine tools.

The importance and problems of external linkages for SMEs

The technological linkages of SMEs are many and varied, and are an important feature of innovative smaller companies. External collaboration provides a means for innovative SMEs to complement and supplement their own in-house efforts. External linkages furthermore provide the possibility of an income stream enabling the extension of in-house R&D; a means of commercializing innovations; access to the 'complementary assets' of marketing and distribution; and the ability of larger firms to deal with legal and regulatory issues. Partnerships with well-known, large firms can also improve a SME's credibility with present and future customers, bankers and staff.

Such collaboration takes a variety of forms and, as we shall see in the next chapter, their management takes considerable effort. The component supplier/manufacturer interface appears to be a particularly fertile ground for technological liaison and exchanges during product development.

One aspect of the linkages of SMEs worthy of note is the problems smaller firms possess in linking with large firms. This was seen in the Beesley and Rothwell study and was also found in a sample of European leading technology SMEs studied for the Industrial Research and Development Advisory Committee of the EC (Dodgson and Rothwell 1987). These highly advanced firms shared the concern over working with larger firms, and showed a similar preference for dealing with other SMEs. The lack of linkages between large and small firms in Europe has also been noted by Dekker (1989). The major reason why such collaborations are not formed is the size of the management problems involved for SMEs.

There is an obvious mismatch between the management and other resources available to a large firm compared to a SME. The process of forming linkages – identifying the need for a linkage, finding a partner, managing the arrangement, and ensuring mutual benefit – puts great strains on SMEs' management. There are high opportunity costs involved in forming a collaboration for a SME given the limited amount of management time available. The time horizon of a SME may be shorter than a larger firm's. The procedures involved in a collaboration may appear unusually bureaucratic for a smaller firm. There may be completely different cultures amongst the managers, scientists and engineers in firms of different sizes. The smaller firm may not be in the most advantageous position concerning any IPRs which may emerge, particularly given the high cost of legal and patenting advice. These, and other management problems for SMEs are discussed in more detail in the next chapter.

For all these reasons, SMEs may choose not to collaborate with large firms. One of the prime stimuli to collaboration is the employment of QSE. The effective use of these employees is related to the quality of management in SMEs, particularly their outward-lookingness and receptivity to external ideas. There is value in SMEs adopting a strategic approach to their external linkages (Dodgson and Rothwell 1991). This was seen particularly clearly in Chapter 5 in the case of Celltech, which has managed very effectively to combine internal and external expertise in a complementary fashion. This strategic perspective proved extremely effective in the studies reported by Mayer-Krahmer and Kuntze (1987) whereby a number of German traditional engineering companies made the transition to the design and manufacture of high technology products through the use of external technology. As will be argued in the following chapter, the strategic skills involved in building long-term and complementary arrangements in these cases apply to large as well as small firms. It is perhaps in the area of the strategic management of external technological linkages that SMEs may find their greatest problems and short-comings.

Public policy and small firm collaboration

In previous chapters it has been seen that while the initial expectation of collaborative programmes and schemes has been to encourage SME participation, this has in practice been subverted over time so that large firms predominate in them. From the small firm's perspective, questions also need to be asked about the relevance of collaborative schemes for smaller firms. Whilst the very small minority of smaller firms which enjoy technological leadership in their areas, and possess marked scientific and technological capabilities, may benefit from public policy promoted collaboration, this does not apply to the majority of innovative small firms. Dickson *et al.* (1990) describe a number of examples of small firms that were 'forced' into inappropriate collaborative schemes, as there was a complete absence of government support for their technological activities on an individual basis (see Chapter 3). Senker (1992) describes the way that small biotechnology firms have found it very difficult to participate in the high technology schemes promoted by the government. Garnsey and Moore (1992) analyse the way that the whole UK policy towards technology during the 1980s has been misguided, particularly in respect to promoting pre-competitive R&D, and has had severe consequences for small firms. More appropriate linkages for smaller firms may be those whose technological aims are not so advanced, and are closer to the market. Few small firms have the financial strength and security to operate with long-term horizons. Their technological demands are therefore much more immediate. The high number of linkages promoted by public policies shown in the Rothwell and Beesley study refer to schemes that then existed promoting individual firm support for innovation. Recently, the mistake made by the British government in relying so extensively on collaborative schemes has been realized, and a number of small schemes have been introduced which provide individual small firms with assistance for innovation. Time will tell whether the government can overcome its ideological objections to intervening 'near market' for small firms.

CONCLUSIONS

This chapter has described the value for small firms in collaborating in their technological activities. Small firms have played a valuable role in the development and diffusion of a number of key technologies. Collaboration is important for the small firm, and can also be important for large firms linking with them, and for the innovation process in general. The extensive external linkages of small firms have been mapped. Some of the constraints facing small firm collaboration have been described, including the shortage of key skilled employees, public policy shortcomings, and concerns over linking with large firms. The poverty of large firm strategy towards collaborating with small firms, thereby missing its potential benefits, will be discussed in the following chapter.

Chapter 12

Management of technological collaboration

The six case studies described in this book have highlighted some of the major management problems in technological collaboration. In this chapter some of the normative management issues of collaboration are summarized. The difficulties in actually establishing the success of collaboration are discussed. Some of the major management issues of partner selection, communications, adpatable structures and trust are analysed. The difficulties of the strategic management of collaboration are revealed by examination of some of the problems large firms have in dealing with small firms.

THE MANAGEMENT PROBLEM WITH INTER-FIRM LINKAGES

It must be borne in mind that a high proportion of a company's R&D efforts are unsuccessful: Ohmae (1990), for example, suggests that this figure is as high as 90 per cent. Yet technological innovation is critical for firms' continuing competitiveness. It is a complex and uncertain process which has *strategic* implications for a firm's long-term performance. So the effective management of innovation is of the utmost priority in attempting to increase the chances of successful R&D. This is particularly so when managers have to address the problems of *integrating* external inputs and carefully and skilfully managing *continuing* linkages with other organizations and firms. If firms did not need to link with other firms they probably wouldn't. Working in partnership with other organizations with different strategies, structures, systems and cultures is a very difficult process and inevitably tension-ridden.

Managers have to think very seriously about partner selection: which firms and organizations can provide what is needed and be trusted to deliver what is expected of them? What project management skills and organization are most effective in encouraging communication? While inter-company linkages are widespread, and potentially can be advantageous, these advantages are usually *hard earned*.

Apart from building up mutual trust between partners, managers need to overcome the suspicion that collaboration is a zero-sum game, and that their firm is giving away its technology. They need to deal with the antipathy towards collaboration which is felt by many scientists and engineers having to work with other organizations. Managers may need to overcome the 'not invented here' syndrome, or to assure staff that collaboration is not a reflection of their lack of skills.

Success and failure in collaboration

It is notoriously difficult to define success in collaboration. The range of firms' circumstances and their expectancies and experiences of collaboration are so variable as to make uniform definitions of success and failure unwise. Some detailed empirical work on assessing outcomes of collaboration has been undertaken in the evaluations of public policies promoting collaboration (see, for example, Guy and Georghiou's (1991) comprehensive analysis of the UK's Alvey Programme). The difficulties of evaluating success and failure are highlighted in this report by the conclusion that the major benefits from the programme were not technological, but more to do with the way it altered the strategies of participants in encouraging collaboration. Other studies show the high level of failure of collaborations. Harrigan (1986), for example, showed that of her sample of 895 joint ventures, only 45 per cent were mutually agreed by their partners to be successful. Dickson *et al.* (1990), in a series of case studies, show the frustrated expectations of firms involved in collaboration, particularly amongst small firms collaborating with large ones. Imai (1990) argues that many Japanese firms have difficulties in joint ventures with foreign firms, and that truly successful cases are few.

Why should there be a high failure rate? Research findings into technological collaboration tend to agree about very little (Dodgson 1991a). One thing that they are in complete agreement about, however, is just how difficult collaborations are to manage (Devlin and Bleackley 1988; Lynch 1990). As Doz (1988) tells us, many partnerships 'do not actually yield the results expected by either or both partners, and even when a measure of success is achieved, the tensions in making partnerships work sometimes dwarf their success in the eyes of the participants' (Doz 1988: 318).

The problems of collaboration relate not only to external components, but to the ways in which the outcomes from collaboration enhance internal capabilities. Linking external inputs into internal R&D work can be a contentious issue. By way of illustration, Lawton Smith *et al.* (1991) provide the following story:

> a small software company had been called into a major telecommunications company to help develop a product which would integrate a computer graphics package and database capability into the existing in-house system. The collaboration included a few senior technical people in the large company, but excluded the development team who had spent two years on the project. The outcome was that successful technical collaboration was throttled because of

major problems within the host organisation. The software company gained
little in financial reward, and the project was eventually dropped.

(Lawton Smith *et al*. 1991: 464)

Arnold, Guy and Dodgson (1992) discuss cases of mixed success, where there is
satisfaction with the outcomes of collaboration amongst some partners, but not
amongst others. The case studies of technological collaboration described in this
book, brief details of which are included in Table 12.1, all enjoyed varying levels
of success. In all cases there was technological success, although in the case of
Quantel this was not immediately joined by market success. The case studies
reflected wide variety in the technologies in which collaboration is commonly
found; size of firm and collaboration; type and length of collaboration; and whether
they were promoted by public policies or not. Despite this variety, senior managers
and project managers in the firms shared common views of the reasons for the
success of the collaboration. These tended to refer to factors such as complementary
technology and trust, respect and good communications between partners. Also
important were management factors, both in project management and with regard
to the *strategic* nature and issues of R&D collaboration. The top managers in the
firms were committed to the objectives of the collaborations and championed their
cause.

However, collaborations between firms show an infinite number of motives,
structures and outcomes. Generalizations about how best to manage collaborations
are, therefore, not appropriate in all cases. As Lynch (1990) argues 'There is no
one correct solution or answer for every alliance; each one must be designed and
managed in its own unique fashion to fit its own circumstances' (Lynch 1990: 22).

The situation is further complicated by the way the aims of collaboration may
alter during the course of the partnership. In a number of the case studies – Racal,
BT&D and Quantel – the outcomes of the collaboration were not the ones initially
expected. The collaborations were, however, no less successful.

If one accepts the distinction between pre-competitive and near-market R&D,
then the tactical, day-to-day management problems facing each may differ. The
strategic perspective in which these tactical considerations arise may remain
common. The innovation process is iterative, and its management should be
integrated throughout its various stages. Strategic management cohesion is necessary throughout the process.

By studying successful collaborations in the case studies a number of features
of the approaches management have adopted can be described. On this basis, and
from a review of previous research, a number of general observations can be made
about the strategic management of collaboration in a number of areas, including
partner selection, adaptability in structures, and communications and human resource issues based on trust.

Table 12.1 Six case studies of technological collaboration

Case study	1 BT&D Technologies Ltd	2 Celltech Group plc	3 Quantel plc	4 GEC Sensors – Tactical Information Division	5 Ricardo Consulting Engineers	6 Racal Research Ltd
Companies	British Telecom Du Pont	Celltech American Cynamid	Quantel SSL	GEC Sensors Alcatel plus others	Ricardo Dalian Locomotive Works (China)	Racal Research Hewlett–Packard Ferranti plus others
Type	Joint venture	Joint R&D project	Joint R&D project	Joint R&D project promoted by Eureka Programme	R&D contract project	Joint R&D project promoted by Alvey Programme
Size	350 employees $100 million initial investment	20 Celltech employees	6 engineers	45 TID employees, £30 million	8 engineers	Up to 45 Racal employees, £7.5 million
Started	1986	1986	1987–9	1990	1980	1984
Technological focus	Optoelectronics	Antibody engineering	Image/sound processing	Terrestrial/in-flight telephone	Engine design and improvement	Mobile information systems

Partner selection

Partner selection is the most critical decision affecting the success of collaboration. The case studies and the literature review reveal the advantages in selecting partners with whom to develop long-term relationships (Buckley and Casson 1988; Dickson *et al.* 1990). As many of the motives for technological collaboration reflect attempts to deal with complexity and uncertainty in novel and rapidly changing technologies, and to transfer knowledge which is tacit and firm-specific, it is perhaps unsurprising that there are advantages in partnerships with long time horizons. In long-term relationships the problems in establishing effective communications may be overcome. There is greater opportunity for firms to exchange information equitably. Managers and engineers in different companies can better develop working relationships. Technological capabilities are more easily and comprehensively transferred. If these advantages are to be obtained, selection of the partner should be made on the basis of the long-term attractiveness of the collaboration, as well as the intrinsic interest of the proposed project. In this respect partner selection should be a strategic decision. Five of the case study companies took great care over partner selection (the sixth company, Quantel, partnered SSL as it was in the same group of companies). For example, Du Pont had spent some years searching for a partner and Celltech and Cyanamid spent a number of years 'getting to know' one another, before expanding their relationship. Negotiation over partner selection was always a function of top management.

As Figure 12.1 shows, in the case studies the partners provided *complementary* technologies. The BT&D joint venture received manufacturing and materials technology from Du Pont and optoelectronics from BT. In its partnership, Celltech provided antibody engineering skills, and Cyanamid provided expertise in toxins, the former allowing targeting of a particular problem in the body (a tumour, for example), the latter the method of treatment. Quantel's core technology is image processing, SSL's is sound processing; the two were planned to combine in a single system. GEC Sensors possesses skills in in-flight communications, Alcatel's expertise lies in terrestial communications. These forms of complementarity in expertise were frequently cited as a reason for the technological success of the collaborations which had enabled partners to learn novel skills. They also, of course, reveal the ways in which the firms were not direct competitors in the focus of the collaboration.

Collaborations exist to transfer knowledge. Once this knowledge has been transferred, the need for the partnership may be assumed to be finished. *However*, as technologies and markets are continually developing, the transferred knowledge may no longer be the most appropriate for changed market conditions. Individual firms continue in their efforts to develop their specialist skills through R&D. For this reason there are advantages in firms collaborating not only on the basis of existing technology, but on the understanding that partners may continue to improve their technological capabilities. These improvements are related to the comparative advantages of individual firms. For example, within Celltech it is

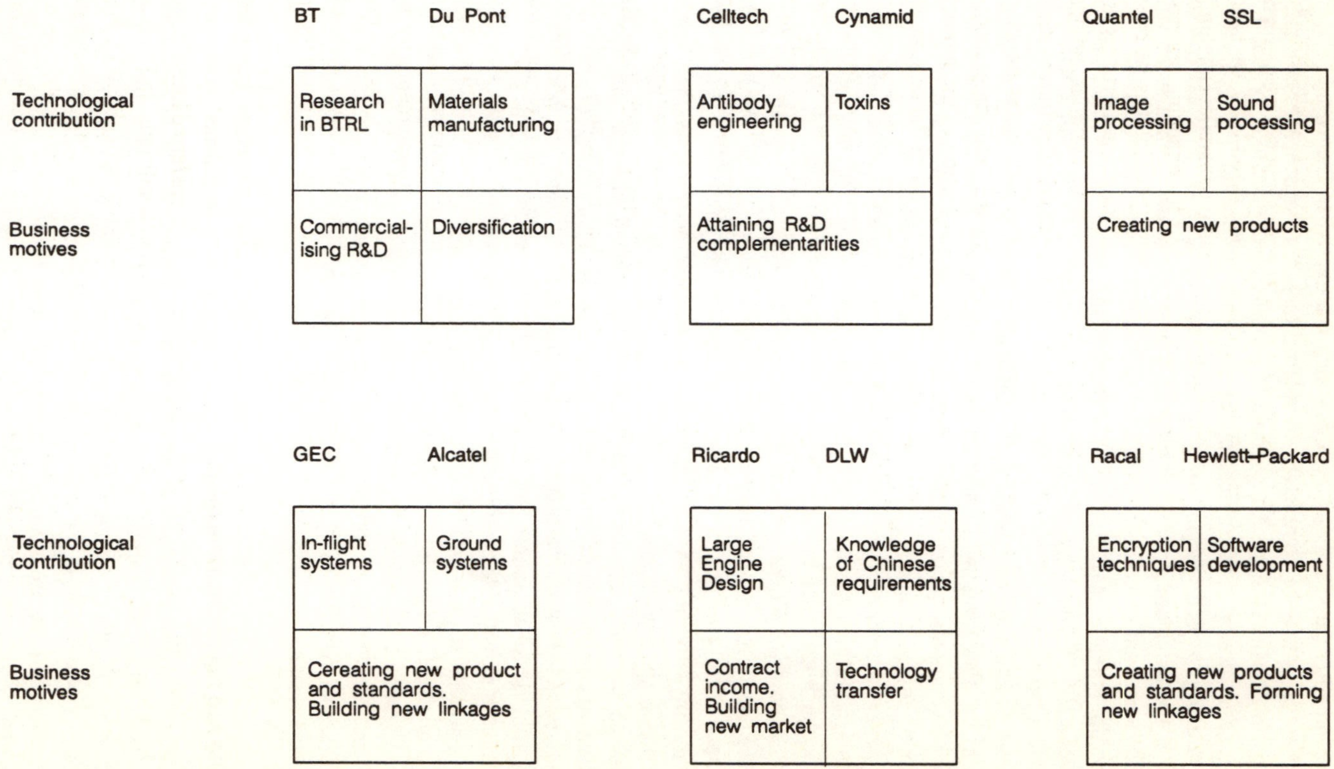

	BT	Du Pont	Celltech	Cynamid	Quantel	SSL
Technological contribution	Research in BTRL	Materials manufacturing	Antibody engineering	Toxins	Image processing	Sound processing
Business motives	Commercialising R&D	Diversification	Attaining R&D complementarities		Creating new products	

	GEC	Alcatel	Ricardo	DLW	Racal	Hewlett-Packard
Technological contribution	In-flight systems	Ground systems	Large Engine Design	Knowledge of Chinese requirements	Encryption techniques	Software development
Business motives	Cereating new product and standards. Building new linkages		Contract income. Building new market	Technology transfer	Creating new products and standards. Forming new linkages	

Figure 12.1 Motives and contributions in the case studies

believed that it is its continuing fast development of expertise which provides its future competitiveness, and that it is this which is attractive to Cyanamid.

Technologies may be completely complementary, but firms may have totally incompatible business aims. An example of complementary technology, but different product strategies, is provided by the case study of Quantel. The technology each partner provided merged together well, but because of different market strategies the collaboration did not work in a business sense. Collaborators will generally not want to compete in the same markets using the product of their partnership. Potential markets need to be demarcated, either on a product basis or geographically. Technology and business strategies need to be sympathetic and mutually supporting.

Scientists and engineers could find it difficult to work with people in similar positions in other firms with lower levels of competence. Specialist vocabulary may not be common, understanding of the latest research techniques or findings may not be shared. Unequal competences result in delays and diversion of efforts as the weaker partner is brought up to speed. Unless there is an element of respect for partner's abilities, transfer is unlikely to be wholehearted. This was found to be of particular importance in the case study of Ricardo Consulting Engineers and Dalian Locomotive Works in China. International collaboration appears to accentuate the need for greater respect for partner's competences. Furthermore, awareness of the commensurate abilities of partners may provide a stimulus to creativity, by providing an element of competitiveness between research teams which may assist innovation.

Tensions between partners can occasionally be reduced with due attention to more tactical considerations before embarking upon a collaboration. For example, decisions need to be made early on in a collaboration about who owns what of any results from the collaboration. Deciding this up-front saves considerable problems when something valuable emerges. Similarly, early agreement on review procedures can save subsequent problems. Review procedures should not be too loose (as projects may drift aimlessly), nor should they be too tight (as these will restrict the ability of projects to change objectives). Agreed project milestones assist the review process. Conditions for termination should be agreed from the start. Getting the balance between flexibility in objectives and their supervised control is a major management task.

Flexible and adaptable structures

The process of collaboration is, as described above, tension-ridden; as Roehl and Truitt (1987) contend 'stormy open marriages are best'. All the case study companies referred to the exceptional difficulties in cooperating with partners. In part these tensions derive from the way technologies and markets constantly change. Unless collaborations are dynamic in nature they may be aiming at a target which has moved. Throughout the course of a collaboration, opportunities may arise which were initially unforeseen, and outcomes from collaboration are often not the ones originally envisaged (Lyles 1988). For these reasons, there need to be elements of adaptability in the organization of collaborations. (For greater consideration of the question of adaptable structures in times of rapid technological change, see Dodgson 1992c).

Amongst some of the case studies there are examples of the benefits of adaptation. Ricardo International and Celltech have progressively become more involved with their partners over time. BT&D initially aimed at a particular market which did not materialize as expected. It managed to adapt its skills and technology to a different market. There are advantages of having this sort of adaptability and flexibility built into the collaboration's systems and structures (Doz and Schuen 1988). In the case of GEC Sensors and Alcatel, the project structure was encouraged to grow organically rather than be burdened with formal systems at an early stage. As the focus of collaboration may change over time, i.e. as the project progresses nearer to the market, then the skills mix of managers, scientists and engineers needs to adapt accordingly.

Trust, communications and human resources

All the case study companies emphasized the importance of good communications to the success of collaboration. Communication is necessary within and between firms. Building effective communications paths into partners is often problematic, particularly for small firms linking into multinationals. Having established the appropriate reporting linkages, the next problem is using them effectively. Reporting unnecessary or poor quality information may reduce the credibility of the whole system (and the collaboration). Without giving away all the knowledge and skills which made one partner attractive to the other, it is important to transfer information which is necessary to make the collaboration work. Sometimes one partner may feel that they are contributing more than the other. In such circumstances a *quid pro quo* is needed. This may be achieved at later stages in the project, or in future projects. It may be achieved formally, through, for example, proportional allocation of intellectual property rights or equity. Or it may be done informally through the trading or exchange of information at the discretion of project managers. Care needs to be taken in maintaining these communications paths when there are changes in personnel in partner firms.

Effective internal communication is, of course, a necessary concomitant. It is important to keep senior management abreast of developments. Having accessed information from partners, its effective use within a firm depends on its diffusion amongst its various parts.

High trust relationships within collaborations are often based on the assumption of continuity and reciprocity between partners (Axelrod 1984; Sako 1992). The case studies, and other in-depth studies of collaboration, place great emphasis on the *personal* factors which enable trust to develop and collaboration to succeed. Communications depend on individuals, and are enhanced by the ability of individuals to be trusted. Managers, scientists and engineers are trusted by their equivalents in other firms to deliver what is expected of them on time. Counterparts are trusted to be honest, and not to impart false or misleading information. Trust is also important when there is imbalance in contributions to the collaboration. The partner may be trusted to rectify the imbalance in the future. Given the way that dependence on personal trust can be adversely affected by labour turnover, or arguments between individuals, stable and continuing trust depends on the establishment of *inter-organizational* trust. This is characterized by a strong community of interest, organizational cultures receptive to external inputs, and shared comprehension of the status and value of collaboration throughout the organizations (Dodgson 1992a).

Given the importance of trust in collaboration, the development of a good reputation for trustworthiness may be an important strategic management tool, inasmuch as a firm would thereby have a wider choice of partners. Lundvall (1988) refers to the way 'trustworthiness' becomes a decisive parameter of competition.

> If a user has a choice between a producer known for low-price and technically advanced products, but also for having a weak record in terms of moral performance, and one well known for trustworthiness, the first will be passed by. This implies limits to opportunistic behaviour. Those limits are reinforced when users pool their information about the reliability of different producers.
>
> (Lundvall 1988: 353)

Pucik (1988a, 1988b) argues that it is the management of human resources which provides the critical factor determining the success of partners within collaborations. Strategic human resource management is an important aspect of collaboration in a number of respects. First, as Devlin and Bleackley (1988), and all the managers in the case studies, argue, collaboration requires very good project managers. Such personnel need to be attracted into the partnerships without, for example, jeopardizing career and pay prospects. In addition to the usual project management skills of overseeing project objectives, budgetary control and man management, collaboration requires particular diplomatic skills. Getting things done in your own organization is often difficult enough, but it requires great diplomacy and powers of persuasion to get things done the way you want them in others. The project manager is also responsible for overcoming the reservations which many scientists and engineers feel about collaboration. Second, given the importance of interpersonal communications in collaboration, the retention of key individuals, managers, scientists and engineers, is crucial. Third, attention to human resource issues can reduce the tensions which sometimes occur in collaborations due to the lack of harmonization between the salaries and conditions of partners.

A particularly important role in collaboration lies in what the innovation literature calls a 'technological gatekeeper' (Allen 1977), and is known in social psychology as a 'boundary spanner' (Michael 1973). The function includes 'scanning, stimulating data-generating activity, monitoring, evaluating data relevance, transmitting information, and facilitating interpersonal intercourse' (Michael 1973:

240). The tolerance of, and support for, such individuals is critical to the success of collaboration.

Successful collaboration depends critically on a shared culture between scientists and engineers. This, as the following, perhaps apocryphal, story suggests, is not easy to achieve. Staff at the competing computer companies IBM and Apple are renowned for the different approaches taken in their appearance. IBM is famed for its legions of blue suits, whilst Apple staff have a relaxed and casual dressing code. A collaboration between the two firms was negotiated at senior manager level, and an initial meeting between the two collaborating research teams was arranged at a neutral venue. Sensitive of the differences between the two firms, IBM's staff agreed to wear their weekend jeans and sweatshirts. The Apple team arrived wearing newly purchased blue suits! This story is illustrative of the number of levels of potential incompatibility between firms, and the difficulties of developing trust and good communications when firm cultures are so very different.

Particular problems of linkages with small firms

Technological linkages can provide both large and small firms with valuable supplements to their own technological activities. They can offer small firms an income stream to enable them to continue to develop in-house R&D, a means of commercializing innovations, and access to the marketing and distribution skills of large firms and to the abilities of larger partners to deal with regulatory and legal problems. Partnerships with well-known, large firms can also improve a small firm's credibility with present and future customers, bankers and staff. As described in the case of biotechnology, large firms can gain through collaboration with small firms as they can provide entrepreneurial stimulus and can adapt quickly and flexibly to rapid technological change.

Many of the problems facing small firms when collaborating with large firms are common to all collaborations, but are just that much more serious for financially-constrained companies, often with limited market strength. The particular problems include:

1   Fear of unwelcome take-over. Large companies in Britain appear to be more disposed towards acquiring small firms than to working collaboratively with them. Acquisition may result in a high level of turnover of key staff, often removing the reason for the acquisition in the first place (Roberts and Garnsey 1992). In their choice of partner, small firms need to assess the previous behaviour of large firms towards acquisition, and the management skills of the large company in working collaboratively with smaller ones.

2   Imbalance in the significance of the linkage. What for a small firm may be a project of strategic significance for its future competitiveness, may be a minor, marginal project for a large firm. Small firms' exposure to risk may, therefore, be disproportionate. Experienced scientists, engineers and managers in small firms may be confronted at project meetings by far more junior staff from the

large firms. Assessment of the likelihood of this occuring is difficult, but a large firm's reputation for collaboration can be evaluated.

3   Communications between the simple management structures and organization in small firms, and their often byzantine and opaque equivalents in large firms is frequently difficult for the smaller firm. Reporting paths are often unclear, and the location of decision-making authority often difficult and lengthy.

4   Small firms cannot bear the high management cost of negotiating the linkage, which may take some time. They cannot afford the high level of legal protection often accorded to large firms by their lawyers. They find the protection of intellectual property rights through patents very expensive. Again, these problems make the risks of collaboration facing smaller firms comparatively higher than for large firms.

5   Gaining information about potential sources of external know-how in the first place. Small firms rarely have the resources to search for and sift information on research and collaborative programmes of relevance. Their ability to do so, and subsequently to integrate anything they have learnt from an external source, depends, as seen in the last chapter, on the skills and qualifications profile of their employees. In this respect, small firms are disadvantaged compared to larger firms.

Many large firms find it difficult to link with small firms in any strategic manner. As Segal, Quince and Wicksteed argue 'few large firms have thought seriously about their links with small firms in the sense that they have a definite policy, established at top level and made known to concerned individuals throughout the organisation' (Segal, Quince and Wicksteed 1988: 67).

This, the authors contend, is a major shortcoming as a strategic relationship is necessary if the small firm is to play a role in product development and/or provide the large firm with a significant input of technology. The way in which the majority of small firm linkages were found not to derive from formal corporate planning processes had three main consequences. First, it meant that any partnership was inevitably peripheral and failed to command widespead support throughout the company. Second, it was critically dependent on being championed by individuals, and was likely to collapse if they were to move. Third, the partnership was narrowly based; it would, for example, be a link only between technical personnel, and would preclude links with the marketing function so valued by many small firms. Although Segal, Quince and Wicksteed found very little example of good practice on the part of large firms' linkages with small firms, they were impressed with the strategy adopted by Philips. The following description of this strategy comes from Segal, Quince and Wicksteed (1988).

Philips launched its strategic alliance programme in 1987 with the aim of building around itself a network of small high technology firms. The programme was driven by the longer-term needs of Philips' corporate strategy, and focussed investments in five main areas:

1 'windows' on technological opportunities: emerging technologies which are parallel or possible alternatives to Philips' key technologies;
2 'greenhouse' opportunities: riskier new ideas in which Philips is interested but does not wish to pursue under its own name;
3 'core supporting businesses': business ideas using Philips technologies but outside mainstream markets or products;
4 'spin-outs': encouraging Philips staff to spin-out of the main company into small companies in a structured manner;
5 business opportunities which are market rather than technology driven but are still in line with long-term goals rather than short-term profits.

Philips is prepared to take a long-term view of its investments in small firms (5–10 years). It is expected that up to 5 per cent of Philips' annual R&D budget will be committed to the programme with an ultimate aim of having $2 to $3 million invested in each of fifty small firms at any one time.

The strategic alliance programme is separate from, and not a replacement for, existing commercial links with small firms through subcontracting and shorter-term commercial investments. The whole programme is run by a small team of eight executives throughout Europe and the USA.

> In practical terms, the programme operates through a series of co-investor venture capital partners to whom day to day monitoring and management is subcontracted. It is largely the job of the co-investor partner to search out target companies and submit business plans to Philips for approval. Analysis is undertaken by the relevant product division which will also provide marketing and technical support. Because investments are made from the central R&D budget, the cost/risk of supporting the projects does not fall on the product divisions. This is seen as an important factor in winning their support in implementing the programme, whilst involving them closely in the process.
> (Segal, Quince and Wicksteed 1988: 193)

The fact that such a strategy is so uncommon is an indictment of the lack of strategic skills on the part of many Western firms. If such companies find it so difficult to develop strategic technological collaborations with small and relatively powerless firms, the problems they have in linking with large firms which have the power to compete with them directly and, perhaps, the ability to take them over, are very much compounded.

CONCLUSIONS

Because of the motives firms have for collaboration, which primarily relate to uncertainties in rapidly changing technologies and markets and the problems of know-how transfer, and the process of collaboration which requires adaptability and trust, cooperation between firms is a strategic issue. This is seen particularly acutely in the question of partner selection. The management of collaboration is very difficult, and this is something which tends to be ignored in some of the literature advocating collaboration as a path to improved technological competitiveness.

This chapter has focussed on some of the strategic management problems of collaboration. Strategic short-comings amongst firms in respect to collaboration is something revealed by the case studies of Dickson *et al.* (1990). The case study firms reported in this book have helped describe some of the major strategic considerations and problems when undertaking collaboration. It is perhaps the necessarily strategic nature of collaboration which has led to such a high failure rate as managers adopt a short-term, tactical approach. Indeed, as well as a decline in the numbers of future collaborations (see Chapter 2), we could also see an increase in their quality, i.e. more closely integrated and long-term linkages.

In Chapter 4, the possibility that collaboration may be a stimulus to organizational *learning* was raised. This will be examined in the following chapter, but here one of the more practical issues of management learning can be raised. Collaboration provides, of course, not only the opportunity to learn from a partner, but to learn about *how* to partner. This applies in all sorts of areas including partner selection and project management. It also teaches companies what can actually be achieved from collaboration. In two of the case studies a lack of experience of collaboration led to unrealistic expectations of collaboration. Senior managers in GEC saw collaboration as a cost-reducing exercise rather than one which enhances product and market scope and technological capabilities. And BT, which was very inexperienced in joint ventures, was unaware, unlike its more experienced partner, of the problems in starting up new ventures such as BT&D. Without the experienced view of Du Pont, BT&D would have suffered and perhaps not survived. Collaboration provides much-needed lessons for managers in novel ways of doing things.

But there is another problem with collaboration: namely overreliance upon it. Collaboration can provide a supplement to internal technological know-how, but it is unwise to see it as an alternative. A recent comparison of attitudes to science and technology of German and UK managers (CEST 1991), although not conclusive, reveals a profound difference within the two groups concerning future sources of technology. For the German managers in-house R&D was by far the most important future source, while it was of minor significance to the British managers. The most important future source for the British managers was to be joint ventures, while this was hardly of any significance to the Germans. Given the comparative success of German industry, this should provide food for thought for the strategies of British firms. It may well be, therefore, that in order to benefit from collaboration, greater strategic management of technology skills than currently exists is required not only in respect to the process of collaboration, but in greater critical awareness of the motives for it and its outcomes.

Chapter 13

The new challenges of technological collaboration

In this chapter a number of questions are raised concerning the future role of collaboration, and what it means for our knowledge of firms and the way they behave. Later sections consider some public policy considerations.

CORPORATE CHALLENGES

One of the major questions for the future is how widespread collaboration will remain. Freeman (1991) posits two, not necessarily contradictory, views of future networking. In the first view

> the upsurge of new networking arrangements is a transitory phenomenon of adaptation to the diffusion of new generic technologies; as firms become more familiar with these technologies they will wish to shift the strategically sensitive areas under their direct and immediate control, i.e. to internalise some of the networks which are now the subject of cooperative arrangements.
>
> (Freeman 1991: 510)

The other view is that networking between autonomous firms will grow in importance and will become the normal method of product and process development. Important in this model is the role of information technology:

> IT not only greatly facilitates various forms of networking, but has inherent characteristics, such as rapid change in design, customisation, flexibility and so forth, which, together with its systemic nature and the variety and complexity of applications, will lead to a permanent shift of industrial structure and behaviour. This will assign to networking a greatly enhanced role in the future.
>
> (Freeman 1991: 511)

In the Freeman/Perez (1988) analysis of changing techno-economic paradigm, institutions and organization forms in firms and society respond to and influence technological change. Each successive techno-economic paradigm is associated with different inputs of large and small firms, management style and practices, and relationships between firms and other organizations. This model, in sympathy with the approaches of Schumpeter on long-wave development, and the product lifecycle approach of Abernathy and Utterback (1975), implies a cyclical role for collaboration based on *uncertainty*. Thus there are periods of high interaction between organizations with numbers of new entrants possessing technological advantages, collaborating widely until a 'dominant design' emerges in a technology. As the technology matures, company efforts focus on production and scale-economies rather than product development, and collaborative activity recedes. Such an approach is allied to the first of Freeman's two views described above.

Allowing for the proposition suggested earlier, that the quantity of collaborations will in future be replaced by their greater quality, and accepting that cost-sharing will remain an important motive for many firms, four key factors suggest that collaboration will continue to remain an important feature within industry. The first relates to the actual role of collaboration, which cannot in any sustainable way be anything but a *supplement* rather than an *alternative* to a firm's core method of technology development; internal R&D. Firms need differential internal competences with which to trade in collaborations. Technology is such a key element of corporate competitiveness that there are obvious pressures to internalize it. As seen in earlier chapters, although some firms are increasing their external R&D this remains a comparatively small element of their total efforts (Haour 1991, for example, suggests that generally less than 10 per cent of firms' R&D budgets are spent externally), and even the huge resources put into collaborative public policy programmes are dwarfed by individual company expenditure. Collaboration is not an either/or option to internal R&D; it is a comparatively small-scale adjunct. Given that in-house R&D efforts are subject to the tendencies of firm-specificity of tacit knowledge and strategies and organizational introspection mentioned earlier, collaboration is likely to remain a potentially very valuable adjunct.

Second, the continuing *complexity* of science and technology, and the *multiplicity* of potential sources of technology, with the possibilities of specialist inputs from, for example, small firms, will ensure continuing *uncertainty*. Scientific knowledge will continue to grow and to be disseminated. Industrial investment in university research continues to increase, and in Japan is increasing very rapidly. The dissemination of scientific and technological knowledge is expanding through the number and range of journals, and through the ways in which it can be accessed electronically. The growing use of the English language in science and business further assists knowledge flow. There will remain an enormous multiplicity of sources of technology. Internationally diverse research groups and specialist firms of all sizes can continue to produce differential and advanced knowledge, perhaps *unexpected* knowledge, which will attract the interest of partners. Small firms will continue to possess the advantages of flexibility and creativity unavailable to large firms and hence provide the basis for continuing collaboration. The large, and increasing, numbers of firms working in high technology supports the view of continuing heterogeneity in sources of technology.

Third, whereas the innovation process has always involved close interaction between firms in clusters – and regional networks of customers and suppliers and interacting competitors and collaborators – this is the first period in which information technology can cement and intensify these linkages. In Rothwell's (1992) sense, the Fifth Generation Process is qualitatively different from what went before. Continuing demand for innovation is assured. Greater wealth in the developed world and some of the developing countries, and better educated consumers, will lead to greater demand for differentiated and novel products. Demand for innovations in health care and environmental technologies will be extensive. Speed-to-market will continue, as it is now, to be a fundamental driver of competitiveness. Management systems and procedures in R&D and manufacturing which can improve the speed of delivery of new products, while maintaining or improving upon quality and price parameters, are likely to retained. Improvements and increased levels of integration are likely in the technologies, like CAD/CAM and expert systems, which have facilitated inter-firm linkages of this nature. Collaboration has played an important role in the Fifth Generation Innovation Process, and it would appear unlikely that such advantages, based again on specialist inputs, would be sacrificed in the face of continuing centrality of innovation for corporate competitiveness.

These reasons suggest that collaboration is a strategic issue. It assists the development of novel skills around core capabilities, and can assist technological diversification. A fourth factor suggesting the continuation of collaboration as a strategic issue is allied to the learning activities of firms. There is now much greater knowledge of the innovation process, and its management and organization is the focus of very considerable interest in the research and teaching activities of universities and business schools. There may exist a greater knowledge of the potentials of external inputs ingrained into the routines of firms through tuition and experience.

The question of collaboration and learning was raised earlier. It was argued that firms face organizational pressures towards introspection and create strategies which favour existing ways of doing things. With continuous and rapid market and technology change it is now common to hear calls from management gurus such as Peter Drucker for firms to change themselves into 'learning organizations' (see also Senge 1990). Argyris and Schon's (1976) conception of different *levels* of learning was described. Collaboration has the potential to encourage higher level learning. It provides the possibilities not only of learning about new technologies, but learning about methods of creating future technologies and of the ways those technologies might affect the existing business. It can teach companies new ways of doing things, and can conceivably alter the nature of the business.

Levinthal and March (1981) refer to three distinct forms of learning based on organizational experience:

> The first is adaptation to search strategies. Organizations attempt to modify their propensities to search for new technologies, as well as their propensities to direct that research towards refinement or innovation, on the basis of experience.

Second, organizations improve their search competences. The greater the experience in looking for refinement (or innovation) in a technology, the greater the efficiency in discovering them. Finally, organizations adapt their aspirations. They learn what to hope for.

(Levinthal and March 1981: 310)

In a sense, the experience of collaboration can affect all these three forms of learning. It can increase the propensity to search, improve the ways in which search is underaken, and assist the creation of realistic expectations.

Learning is essential for firms. It is argued to be a central feature of Japanese new product development. In their study of new product development in five successful Japanese companies, Imai, Nonaka and Takeuchi (1985) found 'an almost fanatical devotion towards learning – both within organizational membership and with outside members of the interorganizational network. To them, learning is something that takes place continuously in a highly adaptive and interactive manner' (Imai, Nonaka and Takeuchi 1985: 353).

However, learning moves beyond the focus of new product development 'The importance to the innovation process of organizational learning and unlearning has been underrated, and the development process itself can be a valuable tool for improving the learning process throughout the firm' (Clark, Hayes and Lorenz 1985: 291).

Given the importance of learning, and the ways in which Japanese firms have used it so effectively in their collaborations (Pucik 1988b), it would appear that the development of a learning approach in companies – which appreciates the value of learning, funds it, and attempts to maximise the returns from it – would be an increasingly important influence on firms' collaborative activities.

As described in the case studies, and shown in a wide range of research, effective collaboration and learning between partners depends on high levels of trust (Buckley and Casson 1988; Jarillo 1988). Lundvall (1988), for example, argues that in order to overcome the inevitable uncertainties in product innovations 'Mutual trust and mutually respected codes of behaviour will normally be necessary' (Lundvall 1988: 52). The advantages of 'obligational contracting relationships' in Sako's (1992) sense, have been described. Similarly, Saxenian (1991) argues that

Silicon Valley firms now view relationships with suppliers more as long-term investments than short-term procurement relationships. They recognise collaboration with suppliers as a way to speed the pace of new product introductions and improve product quality and performance.

(Saxenian 1991: 427)

These firms are argued to exchange sensitive information concerning business plans, sales forecasts and costs, and have a mutual commitment to long-term relationships. This involves 'relationships with suppliers as involving personal and moral commitments which transcend the expectations of simple business relationships' (Saxenian 1991: 428).

Hakansson and Johanson (1988) describe a range of these commitments and bonds:

> Interaction between firms develops over time. It takes time to learn about each other's ways of doing and viewing things and how to interpret each other's acts. Relations are built gradually in a social exchange process through which the parties may come to trust in each other.... Over time, as a consequence of interaction, bonds of various kinds are formed by the parties. There may be technical bonds which are related to the technologies employed by the firms, knowledge bonds related to the parties' knowledge about their business, social bonds in the form of personal confidence, administrative bonds related to the administrative routines and procedures of the firms, and legal bonds in the form of contracts between the firms. These bonds create lasting relationships between the firms.
>
> (Hakansson and Johanson 1988: 373)

Freeman (1991) argues:

> Personal relationships of trust and confidence (and sometimes of fear and obligation) are important both at the formal and informal level.... For this reason cultural factors such as language, educational background, regional loyalties, shared ideologies and experiences and even common leisure interests continue to play an important role in networking.
>
> (Freeman 1991: 503)

Trust, according to Sako (1991), is 'a state of mind, an expectation held by one trading partner about another, that the other will behave in a predictable and mutually acceptable manner' (Sako 1991: 377). She argues that there are different reasons for predictability in behaviour, and this allows three types of trust to be distinguished. 'Contractual trust' exists such that each partner adheres to agreements, and keeps promises. 'Competence trust' concerns the expectation of a trading partner performing his role competently. 'Goodwill trust' refers to mutual expectations of open commitment to each other.

> someone who is worthy of 'goodwill' trust is dependable and can be credited with high discretion, as he can be expected to take initiative while refraining from unfair advantage taking... trading partners are committed to take initiatives (or exercise discretion) to exploit new opportunities over and above what was explicitly promised.
>
> (Sako 1991: 379)

Such high levels of trust often underpin the success of Japanese customer/supplier interactions. If this success is to be replicated in R&D, firms need to become accustomed to the idea of dealing with other firms not only on a highly specified contract-by-contract basis, but that arrangements with particularly important potential suppliers of technology should be continuous and more open-ended. If firms are mutually to benefit from increased collaboration, trust must extend beyond that of expectations of partners to contribute what was contractually obliged of them. The level of trust in the relationship should also encompass unexpected and unsolicited suggestions for partner's benefit in the expectation that in the future they may be reciprocated.

The importance of learning and trust in collaborative relationships detracts from explanations of them which are purely economic and instrumental. Inter-personal relationships, which may be influenced by specific cultures, affect the motivation, process and outcome of collaboration. There is another non-instrumental factor which affects collaboration, and that is *power*. This is not economic or market power, although as seen in Chapter 4, strategic competitive analysis emphasizes the importance of this in collaboration. Instead it is power invested by *organizational* position and *political* considerations. The case studies provided some interesting insights and raised some questions about the role of power in forming collaborations. In the Quantel case study, the authority of Quantel's chairman was enough to override SSL's existing research team, and initiate a joint programme which eventually led to the resignation of the team. In the GEC–Sensors' case, the political negotiation process of the standards body and its international dimension raised interesting questions of power distribution, and how this changes. The question of smaller companies' dependence on larger companies, not only economically but also for their reputation, also raises questions of power: could Celltech survive the termination of its arrangement with Cyanamid? In a field of study where there is a need for much more research, that which examines the question of non-economic power in collaboration is very welcome.

Throughout the book the tensions between firms operating cooperatively and competitively have been mentioned time and again. Perhaps the most pressing strategic management skill to be developed in future collaborations is that which can balance the two, and operate in circumstances when relationships between firms are so complex that both apply. The strategic management skills required in collaboration have been analysed, and it is suggested that those firms which have experience of the process and positive outcomes of vertical collaborations are better situated to apply some of the factors such as high trust and good commnications to their horizontal relationships.

Another important future research question, not considered in this book, is the role of networking and technological linkages *within* firms. The extent of the similarities and differences in the character of intra-firm linkages compared to inter-firm linkages may have some signficance for the continuing study of the organization of large, complex firms.

PUBLIC POLICY CHALLENGES

An extensive range of public policies is used around the world for promoting technological collaboration. Some of the ways, forms and reasons for government support for collaboration have been discussed in Chapters 3, 7 and 10. Assuming, as argued above, that collaboration will continue to be an important feature of industry, a number of questions can be raised concerning the *motives* and the *form* of collaboration from a public policy perspective. As for motivation, technological collaboration can be argued to be anti-competitive, exclusive and reflect a position of weakness rather than strength. There are questions over the form of collaboration related to the appropriateness of the type of research being funded, the way large firms predominate in collaborative programmes, whether the research being publically funded would in any case have been undertaken, the level of pro-activeness of policy makers, and the level of adaptability in collaborative policies.

A variety of motives are suggested to explain why governments are involved in the promotion of technological collaboration. These are usually allied to the aim of improving comparative technological performance: through improving the efficiency of the national technology system, or by preventing the access of foreign firms to domestic technologies and markets. Occasionally it is suggested that international collaboration may be one way of reducing trade frictions by inter-meshing the activities of national firms. However, collaboration has commonly been seen, particularly in the USA, as an anti-competitive tool used to frustrate free international trade. Contractor and Lorange (1988), for example, while generally enthusiastic about the benefits of collaboration, refer to the way joint ventures have been 'forced' on companies 'as options reluctantly undertaken, often under external mandates such as government investment laws or to cross protectionist entry barriers in developing and regulated economies' (Contractor and Lorange 1988: 3).

Collaboration certainly has its exclusionary elements (for firms as well as nations). This exclusion may not only reflect an attempt to keep technologically strong nations at bay, but also to keep weaker nations disadvantaged. In Chapter 9 the high proportion of international collaborations within the triad was noted, along with the very limited number of collaborations amongst the weaker nations. Even within Europe there is a problem of the exclusion of southern European countries within EC collaborative programmes (van Tulder and Junne 1988). As firms in these technologically weaker European nations cannot be awarded grants on the basis of their capabilities compared to northern companies, their governments have demanded *juste retour* for their budget contributions and this has led to some regionally-specific schemes, such as the STRIDE Programme.

At the same time as public policy-promoted collaborative projects may exclude weaker nations and firms, they are themselves generally a response of governments following their perception of comparative international weakness. The original institutions of collective industrial research, the British Research Associations, were formed following a concern over declining competitiveness after the First World War. Japan imported the model to build up its perceived weakness in the technological capabilities of small firms. The model was revived in Western industry as it became clear that Japan was developing a clear technological lead in important sectors. It may well be that the high number of Japanese linkages with

US companies and research institutions in biotechnology (OTA 1991), reflects the comparative weakness of the science and technology of biotechnology in Japan.

The idea that collaboration is a sign of weakness is one also applied to individual firms. Porter (1990) argues:

> Alliances are frequently transitional devices. They proliferate in industries undergoing structural change or escalating competition, where managers fear that they cannot cope. They are a response to uncertainty, and provide comfort that the firm is taking action. Alliances offer initial hope in weaker competitors or companies trying to catch up.
>
> (Porter 1990: 67)

There is also conceivably a case to be made which suggests that the technology which emerges from a collaboration will only be as good as its weakest partner. The technology developed might not be the best possible achievable by individual excellent firms.

Despite these concerns, particular instances of public policy collaborative schemes, as seen in Chapter 7, have been argued to have been important for stimulating key domestic industries. Fransman argues that research cooperation also provides a number of *social* benefits:

1   blending of firm-specific distinctive competences;
2   avoiding duplication in research;
3   improving 'industrial system coherence', that is inter-firm user–producer relationships, by encouraging specialization and information flows;
4   pooling information, and specializing in the collection of information, regarding the technologies concerned;
5   enhancing research competition between the cooperating firms;
6   sharing expensive, non-divisible, equipment (Fransman 1990: 279).

Japan is often touted as an example of a situation where government intervention has successfully promoted collaboration and thereby enhanced industrial success. Some of the ways this has been achieved were discussed in Chapter 10, where it was argued that the Japanese government has played an important role in creating the institutions in which collaboration can occur, and as a 'policeman' mediating continuing collaborative activity between highly competitive firms. This latter role has depended upon the expectancy of *continuity* in public policies promoting collaboration and technological innovation. The advantages of the use of collaboration in Japan are elucidated by Levy and Samuels:

> As resources are pooled and risks are shared, participants in research consortia may be willing to undertake more ambitious technology strategies than if they were working alone. Rapid diffusion of technologies and the avoidance of unnecessary duplication mean that resources may be employed more efficiently. Finally, collaboration permits Japanese industry to enjoy the advantages of size and scale without vitiating competitive pressures. Indeed, it could be argued

that, by creating a 'level technological playing field' throughout the nation's industry, research consortia actually increase competition.

(Levy and Samuels 1991: 143)

However, the role the state has played is argued by some to be supportive rather than pro-active.

the Japanese experience shows that a strong government presence is necessary, it shows just as clearly that the public role must be carefully circumscribed. Above all, publically supported cooperative R&D must take its direction from industry. The Japanese government – small and staffed mostly by generalist civil servants – wisely recognises that it is ill-equipped to play the role of technical director. Thus, many of its programs have been developed on the suggestion of industry. And at the level of the individual research association, industry sets the technical agenda, with little government input.

(Heaton 1988: 38)

Others, however, argue that government has played a more proactive role in determining the direction of collaborative technological efforts (Freeman 1987). The example of the Japanese Protein Engineering Research Institute described in Chapter 10 is an example of a MITI-led initiative. And there are argued to be many others (Levy and Samuels 1991; Fransman 1991).

The extent to which public officials should take the lead in stimulating and directing technological efforts varies, of course, with the behaviour of firms in individual countries. Firms in 'dynamic' countries in Pavitt and Patel's (1988) sense, may not need the level of assistance that is necessary in 'myopic' systems. It is to be remembered that the evaluation of the Alvey Programme in 'myopic' Britain found that firms relied on government subsidy to undertake research which they knew was critical to their continuing survival. In such circumstances, a high level of intervention is perhaps more justified.

It is the way that 'industry sets the agenda' that provides one of the most contentious issues of the form of public policy support and funds for technological collaboration. Van Tulder and Junne (1988) describe the way European programmes are formulated by large firms for large firms. And Peterson (1991) raises suspicion of the way the Eureka programme funnels money to firms, particularly large ones, for R&D they would be undertaking on their own even without government support. Establishing whether this is the case or not, whether there is any additionality, is often difficult to assess. One of the ways in which publically-funded research is directed so that ostensibly it does not directly fund firms' more market-oriented R&D efforts, is to focus collaboration on so-called 'pre-competitive research'. This is not, however, unproblematic.

Basic research, which by definition is pre-competitive, is believed to be a public good. It cannot be appropriated by individual firms protecting it through intellectual property rights. Near-market research, however, is argued to have the possibility of short-term commercial advantage for individual firms (and for this reason, firms are argued not to be interested in near-market collaborations unless markets can clearly be demarcated geographically: there are too obviously going to be winners and losers). It is therefore argued that collaboration promoted by public funds is best concentrated on basic research. However, as was discussed in Chapter 3, the process of innovation is an iterative one, and linear models of basic research leading through to applied research are overly simple. If, as the aim of the policies suggests, innovation is to be promoted, then it is insufficient to support only one element of an intricate and iterative process.

Furthermore, the term pre-competitive has been argued to be politically convenient, as in the case of the EC's ESPRIT programme. Designating it as such provides a block exemption to Community competition policy rules. Within the EC, 'competitive R&D' (undertaken by firms and necessary for the launch of a new product) requires case by case exemption (Sharp 1989). Mytelka (1991) also questions whether a number of ESPRIT projects are pre-competitive. As Fransman puts it:

> If research is to be judged to be worth undertaking for a for-profit company, it must be expected, even with uncertainty, to eventually yield competitive advantage. From this point of view, the notion of pre-competitive in for-profit companies is a contradiction in terms.
>
> (Fransman 1990: 282)

The problems SMEs face in forming external linkages, and the predominance of large firms in shaping public policies, such as Alvey and ESPRIT, has implications for public policy. Collaboration between large and small firms has potential beneficial consequences for technological and industrial development, and its promotion is the focus of attention for public policy makers. There is a pressing need to understand better the interfacial problems and patterns of interactions between SMEs and large firms, particularly the management problems involved (Rothwell and Dodgson 1991). In Chapter 7 some of the public policies promoting collaboration were described, and the inclusion of SMEs in these schemes is often high on the policy agenda. Pre-competitive R&D is, however, inappropriate to the needs of SMEs. As seen in Chapter 11, SMEs enjoy particularly strong near-market interactions involving product development along the vertical supplier–manufacturer–customer chain. Many government policies towards SMEs fail to recognize the importance of these vertical linkages in 'D' rather than 'R', and are instead more appropriate for large, research intensive firms.

Public policy plays a particularly important role in reducing the risk of smaller firm participation in collaboration. It underwrites the negotiation costs of joining in collaborative projects. In some cases, where the intellectual property derived from collaborative programmes is public domain, it thereby allows small firms continuing access to it rather than allowing the possibility of it being monopolized by larger firms. Public policies can encourage information diffusion about collaborative opportunities and improve the capacities of smaller firms to integrate lessons from collaboration through assistance in the employment of qualified scientists and engineers. The justification for public policy involvement in directing and supporting the technological efforts of industry is particularly apparent in smaller firms. They provide the pluralism in sources of technology which has public good benefits in improving innovations, and also in preventing complacency and oligopolistic tendencies in large firms.

Not only are public policies required which are less exclusive, more sensitive to the process of innovation and to the needs of SMEs, greater consideration is required of *delivery* mechanisms for the policies. Reflecting the importance of regional innovation networks and clusters, there is growing consideration within Europe of the ways in which policies can be targeted and delivered by regional authorities. As Bianchi and Bellini (1991) contend:

> Regional governments try to create a favorable local environment for industrial development through direct involvement of all the local public and private actors for the definition of common development programs; these programs have to be supported by the integrated use of complementary policy instruments offered by regional, national and Community authorities to promote industrial development and innovation, such as training programs, export consortia, service centers, common research and application technology projects. Local partnership is then the key issue...
>
> (Bianchi and Bellini 1991: 495)

For these authors the local approach is critical for any policy concerned with SMEs or regional development. Policies at all levels have to be delivered in such a way that they are directed not only at specific production sites, but also take into account the local community and environment. As trust is such an important element in collaboration, and it has a strong regional and cultural specificity, public policies need to reflect this.

Another major problem facing government policies is how to adapt them to changing circumstances. Sharp (1989), for example, argues that collaborative programmes like ESPRIT reflect a transitional period in European electronics during industrial restructuring, and questions whether they are still necessary. Certainly there appears to be considerable value in pursuing flexible policy objectives. Japanese policies have proven particularly adaptable (Heaton 1988). Levy and Samuels (1991) argue 'As Japan moved from technological backwater to world leader in many industries, research consortia, the institutions of collaborative research, were not eliminated but reconfigured and readapted to meet the country's new need' (Levy and Samuels 1991: 127).

Public policies need not only to attempt to direct corporate technology strategies (as in the case of ESPRIT, for example), but also to reflect them. Unless there is a high degree of sympathy between public policies and corporate strategies, then any public programmes will be sub-optimal in their effect. In Chapter 2 it was suggested that the rate of increase in the number of collaborations may be *declining*. And elsewhere it has been suggested that *quantity* of collaborative partnerships may be substituted by greater *quality*. If this is so, then the gamut of public policies designed to promote collaboration, particularly apparent in the UK, may need adjustment. Policies need to be adapted to promote high quality, long-term arrangements instead of their present emphasis on purely increasing their number.

The question of adaptation of public policies is particularly appropriate given the rapidly changing nature of technology. This is perhaps seen most clearly in the most pressing of all current technological challenges: environmental issues and the creation of clean technologies. This applies domestically as firms and research organizations attempt to undertake the expensive R&D necessary to meet growing regulatory demand. This, as the case of CERAM in Chapter 6 showed, often requires collaborative activities for firms that cannot afford the new technologies or find them too novel. It also applies at an international level. Many of the world's environmental problems are global in nature, such as the 'greenhouse effect' and CO_2 emissions. Until international standards and regulations and multi-national technological efforts are directed at this problem, then the likelihood of finding permanent solutions is limited. Technological collaboration, and public policies directing it, is likely to be a necessary pre-condition to finding any solution.

References

Abernathy, W. and Utterback, J. (1975) 'Innovation and the Evolution of Technology in the firm', Cambridge, Mass., Harvard University.

ACOST (1990) 'The Enterprise Challenge: Overcoming Barriers to Growth in Small Firms', London, Advisory Council of Science and Technology, HMSO.

Acs, Z. and Audretch, D. (1990) *Innovation and Small Firms*, Cambridge, Mass., MIT Press.

Adler, P. (1990) 'Shared Learning', *Management Science*, 36, 8: 938–57.

Alic, J. (1990) 'Cooperation in R&D' *Technovation*, 10, 5: 319–32.

Allen, T. (1977) *Managing the Flow of Technology: Technology Transfer and the Dissemination of Technological Information Within the R&D Organization*, Cambridge, Mass., MIT Press.

Angell, C., Collins, G., Jones, A. and Quinn, J. (1984) 'Information Transfer in Engineering and Science', London, Technical Change Centre.

Ansoff, I. (1968) *Corporate Strategy*, Harmondsworth, Penguin.

Aoki, M. (1988) *Information, Incentives and Bargaining in the Japanese Economy*, Cambridge, Cambridge University Press.

Aoki, M. and Rosenberg, N. (1987) 'The Japanese Firm as an Innovating Institution', paper presented at the International Economic Association Roundtable Conference, 'Institutions in a New Democratic Society', Tokyo, 15–17 September.

Appleby, C. and Bessant, J. (1987) 'A Future for Foundries?', Brighton, Brighton Polytechnic/Wolverhampton Polytechnic.

Argyris, C. and Schon, D. (1978) *Organizational Learning*, London, Addison Wesley.

Arnold, E. and Guy, K. (1986) *Parallel Convergence: National Strategies in Information Technology*, London, Pinter Publishers.

Arnold, E., Guy, K. and Dodgson, M. (1992) *Linking for Success*, London, National Economic Development Office.

Arrow, K. (1962) 'The Economic Implications of Learning by Doing', *Review of Economic Studies*, 29, 2: 155–73.

Arora, A. and Gambardella, A. (1990) 'Complementarity and External Linkages: The Strategies of the Large Firms in Biotechnology', *The Journal of Industrial Economics*, 28, 4: 361–79.

Arthur, B. (1990) 'Positive Feedbacks in the Economy', *Scientific American*, February: 80–5.

Axelrod, R. (1984) *The Evolution of Cooperation*, Harmondsworth, Penguin.

Baughn, C. and Osborn, R. (1990) 'The Role of Technology in the Formation and Form of Multinational Cooperative Arrangements', *Journal of High Technology Management Research*, 1, 2: 181–92.

Beesley, M. and Rothwell, R. (1987) 'Small Firm Linkages in the United Kingdom', in R. Rothwell and J. Bessant *Innovation, Adaptation and Growth*, Amsterdam, Elsevier.

Bessant, J. (1988) 'Pushing Boxes or Solving Problems: Some Marketing Issues in the Diffusion of Computer Integrated Manufacturing Innovation', *Journal of Marketing Management*, 3, 3: 352–71.

Bessant, J. (1991) *Managing Advanced Manufacturing Technology: The Challenge of the Fifth Wave*, Oxford, Blackwell.

Bessant, J. and Grunt, M. (1985) *Management and Manufacturing Innovation in the United Kingdom and West Germany*, Aldershot, Gower.

Bianchi, P. and Bellini, N. (1991) 'Public Policies for Local Networks of Innovators' *Research Policy*, 20: 487–97.

Bossard Consultants (1989) 'Contract Research Organizations in the EEC', Brussels, Commission of the European Communities.

Buckley, P. and Casson, M. (1988) 'A Theory of Cooperation in International Business' in F. Contractor and P. Lorange *Cooperative Strategies in International Business*, Lexington, Mass., Lexington Books.

Cabinet Office (1991) *Annual Review of Research and Development Expenditure*, London, HMSO.

Cairnarca, G., Colombo, M. and Mariotti, S. (1992) 'Agreements Between Firms and the Technological Life Cycle Model: Evidence from Information Technologies', *Research Policy*, 21: 45–62.

Carter, C. and Williams, B. (1957) *Industry and Technical Progress: Factors Governing the Speed of Application of Science*, London, Oxford University Press.

Casson, M. (1991) *Global Research Strategy and International Competitiveness*, Blackwell, Oxford.

Chesnais, F. (1988) 'Technical Cooperation Agreements Between Firms', *STI Review*, no. 4, December.

CEST (1991) 'Attitudes to Innovation in Germany and Britain: A Comparison', London, Centre for the Exploitation of Science and Technology.

Ciborra, C. (1989) 'Alliances as Learning Experiences: Cooperation, Competition and Change in the High-Tech Industries', mimeo Institute Theseus and University of Trento.

Ciborra, C. (1991) 'Alliances as Learning Experiences: Cooperation, Competition and Change in High-Tech Industries', in L Mytelka *Strategic Partnerships and the World Economy*, London, Pinter Publishers.

Clark, K. (1989) 'Project Scope and Project Performance: the Effects of Parts Strategy and Supplier Involvement on Product Development, *Management Science*, 35, 10: 1247–63.

Clark, K., Hayes, R. and Lorenz, C. (1987) *The Uneasy Alliance: Managing the Productivity–Technology Dilemma*, Cambridge, Mass., Harvard Business School Press.

Contractor, F. and Lorange, P. (1988) 'Why Should Firms Cooperate? The Strategy and Economics Basis for Cooperative Ventures' in F. Contractor and P. Lorange *Cooperative Strategies in International Business*, Lexington, Mass., Lexington Books.

Cooke, P. and Morgan, K. (1991) 'The Network Paradigm: New Departures in Corporate and Regional Development', Regional Industrial Research Report no. 8, Cardiff, Cardiff University.

Department of Trade and Industry (1991) 'EC R&D: A Guide to European Community Industrial Research and Development Programmes', London, DTI.

Daniels, J. and Magill, S. (1991) 'The Utilisation of International Joint Ventures by United States Firms in High Technology Industries', *Journal of High Technology Management Research*, 2, 1: 113–31.

David, P. (1986) 'Technology Diffusion, Public Policy and Industrial Competitiveness', in R. Landau and N. Rosenberg *The Positive Sum Strategy*, Washington, Academy Press.

DeBresson, C. and Amesse, F. (1991) 'Networks of Innovators: A Review and Intoduction to the Issue', *Research Policy*, 20: 363–79.

Dekker, D. (1989) 'Large Company Involvement with SMEs: A European Survey',

European Commission, *Partnership Between Small and Large Firms*, London, Graham and Trotman.

De Meyer, A. (1992) 'Internationalisation of R&D', paper presented at the Third International Conference in Science and Technology Policy Research, Tokyo, 9–11 March.

Devlin, G. and Bleackley, M. (1988) 'Strategic Alliances – Guidelines for Success', *Long Range Planning*, 21, 5: 18–23.

Dibner, M. (1991) 'Tracking Trends in US Biotechnology', *Bio/Technology*, 9, December.

Dickson, K., Lawton-Smith, H and Smith, S. (1990) 'A Bridge Too Far? Problems and Opportunities in Inter-Firm Research Collaboration', paper presented at the 10th Annual International Conference of Strategic Management Society, 'Strategic Bridging', Stockholm, September.

Dodgson, M. (1989) *Technology Strategy and the Firm: Management and Public Policy*, Harlow, Longman.

Dodgson, M. (1990) 'The Shock of the New: The Formation of Celltech and the British Technology Transfer System', *Industry and Higher Education*, June, 97–104.

Dodgson, M. (1991a) *The Management of Technological Collaboration*, London, Centre for the Exploitation of Science and Technology.

Dodgson, M. (1991b) 'Technological Collaboration and Organisational Learning: A Preliminary Review of Some Key Issues', DRC Discussion Paper no. 85, Science Policy Research Unit, Falmer, University of Sussex.

Dodgson, M. (1991c) 'Technological Learning, Technology Strategy and Competitive Pressures', *British Journal of Management*, 2, 2: 133–49.

Dodgson, M. (1991d) *The Management of Technological Learning*, Berlin, De Gruyter.

Dodgson, M. (1991e) 'Strategic Alignment and Organizational Options in Biotechnology Firms', *Technology Analysis and Strategic Management*, 3, 2: 115–25.

Dodgson, M. (1992a) 'Learning, Trust and Technological Collaboration', *Human Relations*, 46.

Dodgson, M. (1992b) 'The Future for Technological Collaboration', *Futures*, 25, 4: 459–70.

Dodgson, M. (1992c) 'Strategies for Technological Learning: New Forms of Organisational Structure' in J. Marceau *Reworking the World: Organisations, Technologies and Cultures in Comparative Perspective*, Berlin, De Gruyter.

Dodgson, M. and Rothwell, R. (1987) 'Patterns of Growth and R&D Activities in a Sample of Small and Medium-Sized High-Technology Firms in the UK, Denmark, Netherlands and Ireland', Research Report, IRDAC Working Party 3, European Commission.

Dodgson, M. and Rothwell, R. (1989) 'Technology Strategy in Small and Medium-Sized Firms', in M. Dodgson *Technology Strategy and the Firm: Management and Public Policy*, Harlow, Longman.

Dodgson, M. and Rothwell, R. (1991) 'Technology Strategies in Small Firms', *Journal of General Management*, 17, 1: 45–55.

Dore, R. (1973) *British Factory – Japanese Factory*, Berkeley, University of California Press.

Dore, R. (1986) *Taking Japan Seriously*, London, Athlone Press.

Dosi, G. (1988) 'Sources, Procedures, and Microeconomic Effects of Innovation', *Journal of Economic Literature*, 26: 1120–71.

Doz, Y. (1986) *Strategic Management in Multinational Companies*, Oxford, Pergamon Press.

Doz, Y. (1988) 'Technology Partnerships Between Larger and Smaller Firms: Some Critical Issues', in F. Contractor and P. Lorange *Cooperative Strategies in International Business*, Lexington, Mass., Lexington Books.

Doz, Y., Prahalad, C. and Hamel, G. (1989) 'Control, Change, and Flexibility: the Dilemma of Transnational Collaboration', in C. Bartlett, C. Doz and G. Hedlund *Managing the Global Firm*, London, Routledge.

Doz, Y. and Shuen, A. (1988) 'From Intent to Outcome: A Process Framework for Partnerships' paper presented at the Prince Bertil Symposium Corporate and Industry Strategies for Europe, Stockholm, 9–11 November.

Economic Planning Agency (1990) *Keizai Hakusho*, Tokyo, Okurasho.

Fagerberg, J. (1987) 'A Technology-Gap Approach to Why Growth Rates Differ', *Research Policy*, 16: 87–99.

Fiol, C. and Lyles, M. (1985) 'Organisational Learning', *Academy of Management Review*, 10, 4: 803–13.

Fransman, M. (1990) *The Market and Beyond: Cooperation and Competition in Information Technology Development in the Japanese System*, Cambridge, Cambridge University Press.

Fransman, M. and Tanaka, S. (1991) 'The Strengths and Weaknesses of the Japanese Innovation System in Biotechnology', Institute for Japanese–European Technology Studies, Edinburgh, University of Edinburgh.

Freeman, C. (1982) *The Economics of Industrial Innovation*, London, Pinter Publishers.

Freeman, C. (1987) *Technology Policy and Economic Performance: Lessons from Japan*, London, Pinter Publishers.

Freeman, C. (1991) 'Networks of Innovators: A Synthesis of Research Issues', *Research Policy*, 20: 499–514.

Freeman, C. and Perez, C. (1988) 'Structural Crises of Adjustment: Business Cycles and Investment Behaviour', in G. Dosi, C. Freeman, R. Nelson, G. Silverberg, and L. Soete *Technical Change and Economic Theory*, London, Pinter Publishers.

Freeman, J. and Barley, S. (1990) 'The Mutual Organization: A New Form of Cooperation in a High-technology Industry', in R. Loveridge and M. Pitt *The Strategic Management of Technological Innovation*, Chichester, Wiley.

Friar, J. and Horwitch, M. (1986) 'The Emergence of Technology Strategy' in M. Horwitch, *Technology in the Modern Corporation*, New York, Pergamon Press.

Furukawa, K., Teramoto, Y. and Kanda, M. (1990) 'Network Organisation for Inter-Firm R&D Activities: Experiences of Japanese Small Businesses', *International Journal of Technology Management*, 5, 1: 27–40.

Fusfeld, H. and Haklish, C. (1987) 'Collaborative Industrial Research in the US', *Technovation*, 5: 305–15.

Gann, D. (1991) 'Technological Change and the Internationalisation of Construction in Europe' in C. Freeman, M. Sharp and W. Walker *Technology and the Future of Europe*, London, Pinter Publishers.

Garnsey, E. and Moore, I. (1992) 'Pre-Competitive and Near Market Research and Development: Problems for Innovation Policy', Management Studies Group Working Paper, Cambridge, University of Cambridge.

Georghiou, L., Metcalfe, S., Gibbons, M., Ray, T. and Evans, J. (1986) *Post-Innovation Performance*, London, Macmillan.

Gibbons, M. (1990) 'New Rules of the Globalization Game', Conference Report, *Futures*, 22, 9: 973–75.

Gibbons, M. and Johnston, R. (1974) 'The Roles of Science in Technological Innovation', *Research Policy*, 3: 220–42.

Granstrand, O. (1991) 'The Economics of Mul-Tech: A Study of Multi-Technology Corporations in Japan, Sweden and the US', paper presented at the Six Countries Workshop 'Technical Competence and Firm Strategy – Implications for Public Policy', Stockholm, 5–6 November.

Granstrand, O. and Sjolander, S. (1990) 'Managing Innovation in Multi-Technology Corporations', *Research Policy*, 19, 1: 35–60.

Granstrand, O., Oskarsson, C., Sjoberg, N. and Sjolander, S. (1990) 'Business Strategies for

Development/Acquisition of New Technologies', Goteborg, Chalmers University of Technology.

Granstrand, O. and Oskarsson, C. (1991) 'Technology Management in 'Mul-Tech' Corporations', paper presented at PICMET '91, Portland International Conference on Management of Engineering and Technology, 27–31 October, Portland, Oregon.

Graves, A. (1988) 'European Design and Engineering Capabilities: A Continuing Strength?', International Motor Vehicles Programme Working Paper, MIT, Cambridge, Mass.

Guy, K. (1989) 'The UK Information Technology Industry and the Alvey Programme', in M. Dodgson *Technology Strategy and the Firm: Management and Public Policy*, Harlow, Longman.

Guy, K. and Georghiou, L. (1991) *Evaluation of the Alvey Programme for Advanced Information Technology*, London, HMSO.

Hagedoorn, J. and Schakenraad, J. (1990) 'Inter-firm Partnerships and Cooperative Strategies in Core Technologies' in C. Freeman and L. Soete: *New Explorations in the Economics of Technical Change*, London, Pinter Publishers.

Hagedoorn, J. and Schakenraad, J. (1992) 'Leading Companies and Networks of Strategic Alliances in Information Technologies', *Research Policy*, 21, 2: 163–90.

Hakansson, H. and Johanson, J. (1988) 'Formal and Informal Cooperation Strategies in International Industrial Networks', in F. Contractor and P. Lorange *Cooperative Strategies in International Business*, Lexington, Mass., Lexington Books.

Hakansson, H. (1989) *Corporate Technological Behaviour: Cooperation and Networks*, London, Routledge.

Hall, S. (1987) *Invisible Frontiers: The Race to Synthesize a Human Gene*, London, Sidgwick and Jackson.

Hamel, G., Doz, Y. and Prahalad, C. (1989) 'Collaborate with your Competitors – and Win', *Harvard Business Review*, Jan–Feb: 133–9.

Hamilton, W., Vila, J. and Dibner, M. (1990) 'Patterns of Strategic Choice in Emerging Firms: Positioning for Innovation in Biotechnology', *California Management Review*, Spring, 73–86.

Harrigan, K. (1986) *Managing for Joint Venture Success*, Lexington, Mass., Lexington Books.

Harrigan, K. (1987) 'Joint Ventures: A Mechanism for Creating Strategic Change', in A. Pettigrew *The Management of Strategic Change*, Oxford, Blackwell.

Harrigan, K. (1988) 'Strategic Alliances and Partner Assymetries', in F. Contractor and P. Lorange *Cooperative Strategies in International Business*, Lexington, Mass., Lexington Books.

Haour, G. (1991) 'Stretching the Knowledge-base of the Enterprise Through Contract Research', paper presented at the Conference on External Acquisition of Technology, Kiel, 8–10 July.

Hausler, J. (1989) 'Industrieforschung in der Forschungslandschaft der Bundesrepublik ein Datenbericht', MPIFG Discussion Paper 89/1, Cologne, Max Planck Institute für Gesellschafteforschung.

Heaton, G. (1988) 'The Truth About Japan's Cooperative R&D', *Issues in Science and Technology*, Fall, 32–40.

Hedberg, B. (1981) 'How Organizations Learn and Unlearn', in P. Nystrom and W. Starbuck *Handbook of Organizational Design: Vol 1*, Oxford, Oxford University Press.

Hegert, M. and Morris, D. (1988) 'Trends in International Collaborative Agreements', in F. Contractor and P. Lorange *Cooperative Strategies in International Business*, Lexington, Mass., Lexington Books.

Hicks, D., Isard, P. and Hirooka, M. (1992) 'Science in Japanese Companies', mimeo Science Policy Research Unit, Falmer, University of Sussex.

Hladik, K. (1988) 'R&D and International Joint Ventures', in F. Contractor and P. Lorange *Cooperative Strategies in International Business*, Lexington, Mass., Lexington Books.

Hobday, M. (1991) 'Dynamic Networks, Technology Diffusion and Complementary Assets: Explaining US Decline in Semiconductors', DRC Discussion Paper 78, Science Policy Research Unit, Falmer, University of Sussex.

House of Lords Select Committee on Science and Technology (1991) *Innovation in Manufacturing*, London, HMSO.

Howells, J. (1990) 'The Location and Organization of Research and Development: New Horizons', *Research Policy*, 19: 133–46.

Hu, Y.-S. (1992) 'Global or Stateless Corporations are National Firms with International Operations', *California Management Review*, 34, 2: 107–26.

Imai, K.-I., Nonaka, I. and Takeuchi, H. (1985), 'managing the Product Development Process: How Japanese Companies learn and unlearn' in K. Clark, R. Hayes and C. Lorenz *The Uneasy Alliance: Managing the Productivity–Technology Dilemma*, Cambridge, Mass., Harvard Business School Press.

Imai, K.-I. (1990) 'Japanese Business Groups and the Structural Impediments Initiative', in K. Yamamura *Japan's Economic Structure: Should it Change,*? Society for Japanese Studies, Washington DC, University of Washington.

Inzelt, A. and Vincze, J. (1991) 'To Help or Not to Help Asymmetrical Cooperation', paper presented at the conference 'The Role of EC Investment in Promoting R&D Capability and Technological Innovation in Eastern European Countries', Brussels, 19–20 September.

James, G. (1989) *Interfirm Alliances Concerning Technological Developments: An Industry/Technology Lifecycle Perspective*, Canberra., Bureau of Industry Economics.

Jarillo, J. (1988) 'On Strategic Networks', *Strategic Management Journal*, 19: 31–41.

Kelley, M. and Brooks, H. (1990) 'External Learning Opportunities and the Diffusion of Process Innovations to Small Firms', Working Paper 90–15, School of Urban and Public Affairs, Carnegie Mellon University, Pittsburgh.

Kelley, M. (1992) 'Technical, Economic and Organizational Factors Influencing the Propensity to Adopt Programmable Automation Among US Manufacturers', School of Urban and Public Affairs, Carnegie Mellon University, Pittsburgh.

Kleinknecht, A. and Reijnen, J. (1991) 'Why Do Firms Cooperate on R&D? An Empirical Study', mimeo, Amsterdam, University of Amsterdam.

Klepper, S. (1988) 'Collaborations in Robotics', in D. Mowery *International Collaborative Ventures in US Manufacturing*, Cambridge, Mass., Ballinger.

Kodama, F. (1987) 'Japanese Innovation in Mechatronics Technology', *Science and Public Policy*, 13, 1: 291–6.

Kogut, B. (1988) 'Joint Ventures: Theoretical and Empirical Perspectives', *Strategic Management Journal*, 9: 312–32.

Kogut, B. (1988) 'A Study of the Life Cycle of Joint Ventures' in F. Contractor and P. Lorange *Cooperative Strategies in International Business*, Lexington, Mass., Lexington Books.

Komoda, F. (1989) 'The Nature of Technical Linkage and the Logic of Technology Transfer', in H. Singer, N. Hatti and R. Tandon *Joint Ventures and Collaborations*, New Delhi, Indus Publishing.

Koot, W. (1988) 'Underlying Dilemmas in the Management of International Joint Ventures', in F. Contractor and P. Lorange *Cooperative Strategies in International Business*, Lexington, Mass., Lexington Books.

Kreiner, K. and Schultz, M. (1990) 'Crossing the Institutional Divide: Networking in Biotechnology', paper presented at the 10th Annual International Conference of Strategic Management Society, 'Strategic Bridging', Stockholm, September.

Kuhlmann, S. and Kuntze, U. (1991) 'R&D Cooperation by Small and Medium-Sized

Companies', in Proceedings of PICMET '91, Portland International Conference on Management of Engineering and Technology, 27–31 October, Portland, Oregon.

Lamming, R. (1987) 'Towards Best Practice: A Report on Components Supply in the UK Automotive Industry', Brighton, Brighton Business School.

Lamming, R. (1992) 'Supplier Strategies in the Automotive Components Industry: Development Towards Lean Production', unpublished D Phil thesis, Falmer, University of Sussex.

Langlois, R. (1989) 'Economic Change and the Boundaries of the Firm', in B. Carlsson *Industrial Dynamics*, Berlin, Kluwer.

Lash, S. and Urry, J. (1987) *The End of Organised Capitalism*, Cambridge, Polity Press.

Lawton-Smith, H., Dickson, K., and Lloyd Smith, S. (1991) 'There are Two Sides to Every Story: Innovation and Collaboration Within Networks of Large and Small Firms', *Research Policy*, 20: 457–68.

Levinthal, D. and March, J. (1981) 'A Model of Adaptive Organizational Search', *Journal of Economic Behaviour and Organization*, 2: 307–33.

Levitt, B. and March, J. (1988) 'Organisational Learning', *Annual Review of Sociology*, 14: 319–40.

Levy, J. and Samuels, R. (1991) 'Institutions and Innovation: Research Collaboration as Technology Strategy in Japan', in L. Mytelka *Strategic Partnerships and the World Economy*, London, Pinter Publishers.

Lewis, J. (1990) *Partnerships for Profit: Structuring and Managing Strategic Alliances*, New York, Free Press.

Link, A. and Tassey, G. (1987) *Strategies for Technology-Based Competition*, Lexington, Mass., D.C. Heath.

Loveridge, R. and Pitt, M. (1990) *The Strategic Management of Technological Innovation*, Chichester, Wiley.

Lundvall, B. (1988) 'Innovation as an Interactive Process: from User-Producer Interaction to the National System of Innovation', in G. Dosi, C. Freeman, R. Nelson, G. Silverberg and L. Soete *Technical Change and Economic Theory*, London, Pinter Publishers.

Lyles, M. (1988) 'Learning Among Joint Venture-Sophisticated Firms', in F. Contractor and P. Lorange *Cooperative Strategies in International Business*, Lexington, Mass., Lexington Books.

Lynch, R. (1990) 'Principles and Practice of Corporate Alliance Management', *Engineering Management Review*, 18, 3: 22–43.

Lynn, L. (1988) 'Multinational Joint Ventures in the Steel Industry', in D. Mowery *International Collaborative Ventures in US Manufacturing*, Cambridge, Mass., Ballinger.

Macdonald, S. (1992) 'Formal Collaboration and Informal Information Flow', *International Journal of Technology Management*, 7, 1–3: 49–61.

Maidique, M. and Zirger, B. (1985) 'The New Product Learning Cycle', *Research Policy*, 14: 299–313.

March, J. (1991) 'Exploration and Exploitation in Organizational Learning', *Organization Science*, 2, 1: 71–87.

Marceau, J. (1992) *Reworking the World: Organisations, Technologies and Cultures in Comparative Perspective*, Berlin, De Gruyter.

Mariti, P. and Smiley, R. (1983) 'Cooperative Agreements and the Organisation of Industry', *Journal of Industrial Economics*, 31, 4: 437–51.

Maschlup, F. (1982) *Knowledge: Its Creation, Distribution and Economic Significance*, Princeton, Princeton University Press.

Mayer-Krahmer, F. and Kunze, U. (1987) 'Technology Intensive Small and Medium-Sized Enterprises', Report to IRDAC Working Party 3, European Commission, Brussels.

Michael, D. (1973) *On Learning to Plan – and Planning to Learn*, San Francisco, Jossey–Bass.

Miles, R. and Snow, C. (1986) 'Organizations: New Concepts for New Forms', *California Management Review*, 27, 3: 62–73.

Mody, A. (1989) 'Changing Firm Boundaries: Analysis of Technology-Sharing Alliances', Industry Series Paper no 3, Washington, The World Bank.

Mody, A. (1990) 'Learning Through Alliances', Washington, The World Bank.

Morgan, G. (1986) *Images of Organization*, Beverly Hills, Sage.

Mowery, D. (1981) 'The Emergence and Growth of Industrial Research in American Manufacturing, 1899–1946', mimeo, Stanford, Stanford University.

Mowery, D. (1987) *Alliance Politics and Economics: Multinational Joint Ventures in Commercial Aircraft*, Cambridge, Mass., Ballinger.

Mowery, D. (1988) *International Collaborative Ventures in US Manufacturing*, Cambridge, Mass., Ballinger.

Moxon, R., Roehl, T. and Truitt, J. (1988) 'International Cooperative Ventures in the Commercial Aircraft Industry: Gains, Sure, But What's My Share?', in F. Contractor and P. Lorange *Cooperative Strategies in International Business*, Lexington, Mass., Lexington Books.

Mytelka, L. (1991) 'Crisis, Technological Change and the Strategic Alliance' in L. Mytelka *Strategic Partnerships and the World Economy*, London, Pinter Publishers.

Mytelka, L. (1991) 'States, Strategic Alliances and International Oligopolies: The European ESPRIT Programme', in L. Mytelka *Strategic Partnerships and the World Economy*, London, Pinter Publishers.

National Science Foundation (1987) *Science and Engineering Indicators*, Washington, DC, National Science Board.

National Science Foundation (1989) *Science and Engineering Indicators*, Washington, DC, National Science Board.

National Science Foundation (1991) *International Science and Technology Update: 1991*, Washington, DC, National Science Foundation.

Narin, F. and Noma, E. (1985) 'Is Technology Becoming Science?', *Scientometrics*, 7, 3–6: 369–81.

NEDO (1986) *Corporate Venturing: A Strategy for Innovation and Growth*, London, NEDO.

Nelson, R. and Winter, S. (1982) *An Evolutionary Theory of Economic Change*, Cambridge, Mass., Belknap Press.

Nelson, R. (1988) 'Institutions Supporting Technical Change in the United States', in G. Dosi, C. Freeman, R. Nelson, G. Silverberg and L. Soete *Technical Change and Economic Theory*, London, Pinter Publishers.

Nueno, P. and Oosterveld, J. (1988) 'Managing Technological Alliances', *Long Range Planning*, 21, 3: 11–17.

OECD (1989) *Biotechnology: Economic and Wider Impacts*, Paris, OECD.

OTA (1991) *Biotechnology in a Global Economy*, Washington, DC, Office of Technology Assessment.

Ohmae, K. (1989) 'The Global Logic of Strategic Alliances', *Harvard Business Review*, March–April: 143–54.

Ohmae, K. (1990) *The Borderless World: Power and Strategy in the Interlinked Economy*, London, Collins.

Osborn, R. and Baughn, C. (1990) 'Forms of Interorganizational Governance for Multinational Alliances', *Academy of Management Journal*, 33, 3: 503–19.

Ouchi, W. (1986) *The M-Form Society*, New York, Avon.

Ouchi, W. and Bolton, M. (1988) 'The Logic of Joint Research and Development', *California Management Review*, 30, 3: 9–33.

Parolini, C. (1990) 'Growth Paths for Small and Medium High Tech Companies', paper presented at the Symposium on Growth and Development of Small High Tech Businesses, Cranfield Institute of Technology, April.

Patel, P. and Pavitt, K. (1991a) 'Europe's Technological Performance', in C. Freeman, M. Sharp and W. Walker *Technology and the Future of Europe*, London, Pinter Publishers.

Patel, P. and Pavitt, K. (1991b) 'Large Firms in the Production of the World's Technology: an Important Case of Non-Globalisation', *Journal of International Business Studies*, 22, 1: 1–21.

Pavitt, K. (1987) Comment on M. Tushman and P. Anderson 'Technological Discontinuities and Organizational Environments', in A. Pettigrew *The Management of Strategic Change*, Oxford, Blackwell.

Pavitt, K. (1988) 'International Patterns of Technological Accumulation' in N. Hood and J.-E. Vahlne *Strategies in Global Competition*, London, Croom Helm.

Pavitt, K. (1991) 'Key Characteristics of the Large Innovating Firm', *British Journal of Management*, 2: 41–50.

Pavitt, K. and Soete, L. (1980) 'Innovative Activities and Export Shares: Some Comparisons Between Industries and Countries', in K. Pavitt *Technical Innovation and British Economic Performance*, London, Macmillan.

Pavitt, K. and Patel, P. (1988) 'International Distributions and Determinants of Technological Activities', *Oxford Review of Economic Policy*, 4, 4: 35–55.

Pavitt, K., Robson, M. and Townsend, J. (1989) 'Accumulation, Diversification and Organisation of Technological Activities in UK Companies, 1945–83', in M. Dodgson *Technology Strategy and the Firm*, Harlow, Longman.

Peterson, J. (1990) 'Assessing the Performance of European Collaborative R&D Policy: The Case of Eureka', mimeo, University of York, Department of Politics.

Piore, M. and Sabel, C. (1984) *The Second Industrial Divide*, New York, Basic Books.

Pisano, G. (1989) 'The Governance of Innovation: Vertical Integration, Joint Ventures, and Licensing in the Biotechnology Industry', Harvard Business School.

Pisano, G., Russo, M. and Teece, D. (1988) 'Joint Ventures and Collaborative Arrangements in the Telecommunications Equipment Industry', in D. Mowery *International Collaborative Ventures in US Manufacturing*, Cambridge, Mass., Ballinger.

Pisano, G., Shan, W. and Teece, D. (1988) 'Joint Ventures and Collaboration in the Biotechnology Industry', in D. Mowery *International Collaborative Ventures in US Manufacturing*, Cambridge, Mass., Ballinger.

Polanyi, M. (1962) *Personal Knowledge: Towards a Post-Critical Philosophy*, New York, Harper and Row.

Porter, M. and Fuller, K. (1986) 'Coalitions and Corporate Strategy', in M. Porter *Competition in Global Industries*, Boston Press, Boston, Harvard Business School.

Porter, M. (1987) 'From Competitive Advantage to Corporate Strategy', Harvard Business Review, May–June: 43–59.

Porter, M. (1990) *The Competitive Advantage of Nations*, New York, Free Press.

Prahalad, C. and Hamel, G. (1990) 'The Core Competence of the Corporation', *Harvard Business Review*, May–June: 79–91.

Pucik, V. (1988a) 'Strategic Alliances, Organisational Learning, and Competitive Advantage: the HRM Agenda', *Human Resource Management*, 27, 1: 77–93.

Pucik, V. (1988b) 'Strategic Alliances with the Japanese: Implications for Human Resource Management', in F. Contractor and P. Lorange *Cooperative Strategies in International Business*, Lexington, Mass., Lexington Books.

Reddy, N. (1987) 'Voluntary Products Standards: Linking Technical Criteria to Marketing Decisions', *IEEE Transactions on Engineering Management*, 34, 4: 236–43.

Reddy, N. (1990) 'Product Self-Regulation', *Technological Forecasting and Social Change*, 38: 49–63.

Reddy, N., Cort, S. and Lambert, D. (1989) 'Industrywide Technical Product Standards', *R&D Management*, 19, 1: 13–25.

Ringe, M. (1991) 'The Contract Research Business in the UK', Science and Engineering Policy Studies Unit, Policy Study no. 6, London.

Roberts, E. (1991) *Entrepreneurs in High Technology: Lessons from MIT and Beyond*, Oxford, Oxford University Press.

Roberts, E. and Berry, C. (1985) 'Entering New Business: Strategies for Success', *Sloan Management Review*, vol. 26, no. 3, published in E. Roberts (1987) *Generating Technological Innovation*, Oxford, Oxford University Press.

Roberts, J. and Garnsey, E. (1992) 'Acquisition and the Integration of New Ventures; Technology and Culture', in J. Marceau *Reworking the World: Organisations, Technologies and Cultures in Comparative Perspective*, Berlin, De Gruyter.

Roehl, T. and Truitt, J. (1987) 'Stormy Open Marriages are Better: Evidence From US, Japanese and French Cooperative Ventures in Commercial Aircraft', *Columbia Journal of World Business*, Summer: 87–95.

Rosenberg, N. (1982) *Inside the Black Box: Technology and Economics*, Cambridge, Cambridge University Press.

Rothwell, R. (1976) 'Innovation in Textile Machinery: the Results of a Postal Questionnaire', *R&D Management*, 6, 3: 131–8.

Rothwell, R. (1983) 'Innovation and Firm Size: A Case of Dynamic Complementarity', *Journal of General Management*, 8, 6: 5–25.

Rothwell, R. (1986) 'Innovation and Re-innovation: A Role for the User', *Journal of Marketing Management*, 2, 2: 109–23.

Rothwell, R. (1989a) 'Small Firms, Innovation and Industrial Change', *Small Business Economics*, 1, 1: 51–64

Rothwell, R. (1989b) 'SMEs, Inter-Firm Relationships and Technological Change', *Entrepreneurship and Regional Development*, 1: 275–91.

Rothwell, R. (1992) 'Successful Industrial Innovation: Critical Factors for the 1990s', *R&D Management*, 22, 3.

Rothwell, R. and Beesley, M. (1987) 'Pattern of External Linkages of Innovative Small and Medium-Sized Firms in the United Kingdom', *Piccola Impressa*, 2: 15–32.

Rothwell, R. and Beesley, M. (1989) 'The Importance of Technology Transfer', in J. Barber, S. Metcalfe and M. Porteous *Barriers to Growth in Small Firms*, London, Routledge.

Rothwell, R., Freeman, C., Horley, A., Jervis, V., Robertson, Z. and Townsend, J. (1974) 'SAPPHO updated – Project SAPPHO, Phase II, *Research Policy* 3: 258–91.

Rothwell, R. and Zegveld, W. (1985) *Reindustrialization and Technology*, London, Longman.

Rothwell, R., Dodgson, M. and Lowe, S. (1989) *The Technology Transfer System in the UK and Leading Competitor Nations*, London, National Economic Development Office.

Rothwell, R. and Dodgson, M. (1991) 'External Linkages and Innovation in Small and Medium-Sized Enterprises', *R&D Management*, 21, 2: 125–37.

Rothwell, R. and Dodgson, M. (1992) 'European Technology Policy Evolution: Convergence Towards SMEs and Regional Technology Transfer', *Technovation* 12, 4: 223–38.

Sako, M. (1991) 'The Role of 'Trust' in Japanese Buyer–Suppier Relationships', *Ricerche Economiche*, 45 2–3: 375–99.

Sako, M. (1992) *Prices, Quality and Trust: How Japanese and British Companies Manage Buyer Supplier Relations*, Cambridge, Cambridge University Press.

Saxenian, A. (1991) 'The Origins and Dynamics of Production Networks in Silicon Valley', *Research Policy*, 20: 423–37.

Scott-Kemmis, D., Darling, T., Johnston, R., Collyer, F. and Cliff, C. (1990) 'Strategic Alliances in the Internationalisation of Australian Industry', Canberra, Australian Government Publishing Service.

Segal, Quince and Wicksteed. (1988) 'Strategic Partnering and Local Employment Initiatives', Luxembourg, Commission of the European Community.

Senker, J. (1986) 'Technological Cooperation Between Manufacturers and Retailers to Meet Market Demand', *Food Marketing*, 2, 3: 88–100.

Senker, J. and Sharp, M. (1988) 'The Biotechnology Directorate of the SERC: Report and Evaluation of its Achievements 1981–1987', Science Policy Research Unit, Falmer, University of Sussex.

Senker, J. (1992) 'Barriers to Technology Transfer in Small British Biotechnology Firms', Science Policy Research Unit, Falmer, University of Sussex.

Senge, P. (1990) 'The Leader's New Work: Building Learning Organizations', *Sloan Management Review*, 32, 1: 7–23.

Sharp, M. (1989) 'Corporate Strategies and Collaboration: the Case of ESPRIT and European Electronics', in M. Dodgson *Technology Strategy and the Firm: Management and Public Policy*, Harlow, Longman.

Shaw, B. (1988) 'Gaining Value Added from Centres of Excellence in the UK Medical Industry', *R&D Management*, 18, 2.

Simon, H. (1961) Administrative Behaviour. (second edn), New York, Macmillan.

Simon, H. (1991) 'Bounded Rationality and Organizational Learning', *Organization Science*, 2, 1: 125–34.

Stenberg, L. (1990) 'Molecular Beam Epitaxy: A Mesoview of Japanese Research Organization', Mass., Centre for International Studies, Massachusetts Institute of Technology.

Storper, M. and Harrison, B. (1991) 'Flexibility, Hierarchy and Regional Development: the Changing Structure of Industrial Production Systems and their Forms of Governance in the 1990s', *Research Policy*, 20: 407–22.

Soete, L. (1991) 'Policy Synthesis', OECD Technology Economy Programme, MERIT, University of Limburg, Maastricht.

Spalding, B. (1991) 'Biotech Research Alliances on the Rise', *Bio/technology*, 9, December.

Steinmuller, W. (1988) 'International Joint Ventures in the Integrated Circuit Industry', in D. Mowery *International Collaborative Ventures in US Manufacturing*, Cambridge, Mass., Ballinger.

Teece, D. (1986) 'Profiting from Technological Innovation: Implications for Integration, Collaboration, Licensing and Public Policy', *Research Policy*, 15: 285–305.

Teece, D. (1988) 'Technological Change and the Nature of the Firm', in G. Dosi, C. Freeman, R. Nelson, G. Silverberg and L. Soete *Technical Change and Economic Theory*, London, Pinter Publishers.

Teece, D. and Pisano, G. (1987) 'Collaborative Arrangements and Technology Strategy', School of Business Administration, Berkeley, University of California.

Teece, D., Pisano, G. and Schuen, A. (1990) 'Firm Capabilities, Resources, and the Concept of Strategy', CCC Working Paper no 90–8, Berkeley, University of Berkeley.

Thomas, L. (1988) 'Multifirm Strategies in the US Pharmaceutical Industry', in D. Mowery *International Collaborative Ventures in US Manufacturing*, Cambridge, Mass., Ballinger.

Tiler, C. and Gibbons, M. (1990) 'A Case Study of Organisational Learning – The Teaching Company Scheme', *Industry and Higher Education*, March: 47–55.

Tushman, M., Virany, B. and Romanelli, E. (1986) 'Executive Succession, Strategic Reorientation, and Organisational Evolution', in M. Horwitch *Technology in the Modern Corporation: A Strategic Perspective*, New York, Pergamon.

Tushman, M. and Anderson, P. (1987) 'Technological Discontinuities and Organizational Environments', in A. Pettigrew *The Management of Strategic Change*, Oxford, Blackwell.

van Tulder, R. and Junne, G. (1988) *European Multinationals in Core Technologies*, Chichester, Wiley.

von Hippel, E. (1976) 'The Dominant Role of the User in the Scientific Instrument Process', *Research Policy*, 5, 4: 212–39.

von Hippel, E. (1988) *The Sources of Innovation*, Oxford, Oxford University Press.

Walker, G. (1988) 'Network Analysis for Cooperative Interfirm Agreements', in F. Contractor and P. Lorange *Cooperative Strategies in International Business*, Lexington, Mass., Lexington Books.

Werner, J. and Bremer, J. (1991) 'Hard Lessons in Cooperative Research', *Issues in Science and Technology*, Spring: 44–9.

Westney, D. (1988) 'Domestic and Foreign Learning Curves in Managing International Cooperative Strategies', in F. Contractor and P. Lorange *Cooperative Strategies in International Business*, Lexington, Mass., Lexington Books.

Williamson, O. (1975) *Markets and Hierarchies: Analysis and Anti-trust Implications*, New York, Free Press.

Williamson, O. (1985) *The Economic Institutions of Capitalism*, New York, Free Press.

Womack, J. (1988) 'Multinational Joint Ventures in Motor Vehicles', in D. Mowery *International Collaborative Ventures in US Manufacturing*, Cambridge, Mass., Ballinger.

Womack, J., Jones, D. and Roos, D. (1991) *The Machine that Changed the World*, New York, Rawton.

Yarrow, D. (1988) 'US Biotechnology in 1988 – A Review of Current Trends', London, DTI.

Index